MAJOR INDUSTRIAL HAZARDS

Major Industrial Hazards

Their Appraisal and Control

John Withers

HALSTED PRESS
a division of JOHN WILEY & SONS, Inc.
605 Third Avenue, New York, N.Y. 10158
New York • Toronto

Published in the U.S.A. and Canada
by Halsted Press, a division of
John Wiley & Sons, Inc., New York

Library of Congress Catalog Card Number:

ISBN 0-470-21067-2

Contents

List of illustrations vii

Introduction xi

1 Risk analyses and risk perceptions 1
Individual risk – acceptable risk – thresholds and uncertainty – realism and conservatism – societal risk – a structure for risk assessment – public perception of risk – references

2 A classification of major industrial risks with some case histories 23
Explosive, fire, toxic and nuclear hazards – on- and off-site hazards including transport – product hazards and waste disposal problems – references

3 Public policy and legislation 48
The United Kingdom's dual system – the NIHSS Regulations – the CIMAH Regulations – public inquiries – policy positions of public authorities – references

4 Management at all levels 64
Participative control – compliance with codes, procedures and regulations – a manufacturing requirement: the ASME Code – management under the ASME Code – a process management checklist – references

5 Quantifying the release – how big is a hole? 79
'Hazop' and 'hazan' – compensation for under-reporting – the Davenport list – the Kletz list – the Fawcett list – Cremer and Warner studies – top event frequency estimation – references

6 Quantifying dispersion – how long is a piece of string? 99
The concept of equivalent mass – neutral density modelling – dense gas dispersion models – a downwind scaling law – values for the factor *a* – LNG – propane – ammonia – chlorine – butane – hydrogen fluoride – summary and conclusion – references

7 The chances of fire and explosion 127
Illustrations of the relative contributions considered so far – references

8 The damage relationships 138
Damage from explosions – damage from fire and radiant heat – nuclear radiation damage – toxicity relationships – references

9 Assessing the impact upon a local population 168
Methodologies of risk estimates – event trees – rapid methods of hazard assessment – population density – population composition – discussion – references

10 Transport risks 188
Hypothetical example of risks from hazardous road transport – precautionary measures – miscellaneous information – references

11 The mitigation of hazards 200
Design and construction procedures – maintenance – education and training – emergency plans – references

12 The costs and benefits of risk prevention 224
Standard costings – capital cost structure – benefits to employees – the cost of saving a life – benefits to society – references

13 Conclusion 237

Index 243

Illustrations

Figures

1.1 A typical risk transect 6
1.2 Lethal toxicity of ammonia, linear plot 8
1.3 Lethal toxicity of ammonia, log-probability plot 9
1.4 Lethal toxicity of ammonia, LC_{50} values 10
1.5 Risk criteria provided by the Provincial Waterstaat, Groningen 17
2.1 San Juan Ixhuatepec street plan 27
4.1 ASME certificate of authorization 73
5.1 A simple fault tree 82
5.2 Full and observed distribution for the size of an accident 85
5.3 Transformation of $N(z)$ and of z 86
5.4 Frequency/magnitude plots of Davenport data 89
5.5 Frequency/magnitude plots of Kletz data 91
5.6 Comparison of the six 'slopes' 96
6.1 Marshall's equivalent mass relationships 103
6.2 Warren Springs wind tunnel results 113
6.3 Bulk Richardson number versus downwind range 114
6.4 Variability in repeat runs in Warren Springs wind tunnel 117
7.1 Off-site fatalities from on-site explosions 134
7.2 Off-site deaths per annum 135
7.3 Risk transect 136

8.1 Overpressure as a function of explosive mass and distance 141
8.2 Impulse and duration time for 50 per cent lethality 144
8.3 Point source explosions primary blast deaths 50 per cent fatality to 70 kg man 145
8.4 Percentage fatality versus overpressure 146
8.5 Primary and secondary reference lines (blast damage relationships) 148
8.6 Individual risk estimates compared 158
8.7 Societal risk estimates compared 159
10.1 The Hazchem warning sign 195
11.1 The number of tests necessary to establish system reliability 208
11.2 Number of parts and system reliability 209
11.3 A mortality curve: failure rate versus time 210
11.4 The effect of repair time on lost production 216

Tables

1.1 Actuarial data 4
1.2 Risk reports on extra chances of death 5
1.3 Expected mortalities in groups of experimental animals, at 95 per cent confidence level 9
1.4 Differing perspectives 18
1.5 Mortality data 19
2.1 Concentration contrasts between chlorine and ammonia 29
2.2 Lethal toxicity levels for some industrial gases 31
2.3 Predicted Rijnmond annual fatalities 41
3.1 A structure for risk assessment 57
4.1 Perspectives in management 67
4.2 A process management checklist 77
5.1 A hazop guide word checklist 81
5.2 Hazop entry 81
5.3 Frequency/magnitude data based upon Davenport 88
5.4 Frequency/magnitude data based upon Kletz 90
5.5 Frequency/magnitude data based upon Fawcett 92
5.6 Frequency/magnitude data for propylene (Rijnmond) 93
5.7 Frequency/magnitude data for chlorine (Rijnmond) 94

5.8 Frequency/magnitude data for ammonia (Rijnmond) 95
6.1 Estimated rates, masses, TNT equivalents, and yields for actual vapour cloud explosions 104
6.2 Pasquill stability and lateral spread 106
6.3 Neutral density diffusion coefficients 107
6.4 Various model outcomes 111
6.5 Warren Springs wind tunnel results 112
6.6 Field trial test results 116
6.7 Field tests plus models regression analysis 116
6.8 Weather probabilities 119
6.9 Initial conditions for gas dispersion scaling law 120
6.10 Some downwind ranges for chlorine and ammonia 124
7.1 Ignition and explosion data after Wiekema 128
7.2 On-site chances of ignition and explosion 129
7.3 Off-site chances of ignition and explosion 129
7.4 Off-site fatalities from on-site explosions 132–3
8.1 Lovelace field experiments 144
8.2 Explosions at random in populated area (4000 persons per square kilometre) 149
8.3 Some suggested values for TNT equivalence 150
8.4 Comparative hazard ranges 154
8.5 Lifetime risks of premature death 155
8.6 Origin of average annual dose in the UK 156
8.7 Comparative anatomical data 163
8.8 Comparative respiratory data 163
8.9 Lethal concentrations of chlorine (ppm) at an exposure period of 30 minutes 165
9.1 Illustrative probit parameters 169
9.2 Population data after Abercrombie 175
9.3 Population density contrasts 177
9.4 Canvey enumeration district data 181
9.5 A household composition model 181
9.6 Vulnerability data 182
10.1 Comparative transport accident rates 189
10.2 Individual risk comparison 190
10.3 UK hazardous traffic movements 190
11.1 Consultation distances 222
12.1 Illustrative comparison of standard costings 225
12.2 A capital cost estimate 227
12.3 Costs of safety and environmental protection 229

12.4 Fatal accident frequency rate and loss of expectation of life 231
12.5 Delayed deaths 232

Introduction

In June 1974 a devastating explosion occurred at the Nypro works in Flixborough, Lincolnshire, UK. It attracted widespread interest and response from the media and public generally, and stimulated much detailed investigation, both by official inquiry and by the engineering profession. This accident, together with others, is described more fully in Chapter 2. After Flixborough, many papers on the assessment and mitigation of risk were debated in appropriate institutions and learned societies, whilst at national level the UK government set up an Advisory Committee on Major Hazards, the work of which culminated in new legislation.[1,2]

In 1978 the UK Health and Safety Executive (HSE) published a report describing potential hazards in the area of Canvey Island, Essex, UK. Its pioneering work on the methodology of risk assessment aroused international expert interest. This work had been executed for the HSE by the Safety and Reliability Directorate of the UK Atomic Energy Authority, which had previously established its credentials in connection with the assessment of nuclear risks. The Canvey investigations were to continue over a number of years,[3] a second HSE report being issued in 1981. Meanwhile, concern in the Netherlands about risks from major industrial installations in the Europort area led to a comprehensive study by international consultants. Their report to the Rijnmond Public Authority was published in 1982.[4] This provided a further milestone in the development of a practical methodology for risk assessment by experts.

In 1976 a further dramatic accident occurred in Seveso, Italy, which showed that industrial accidents could have serious consequences beyond national frontiers and could cause long-term contaminations which were analogous in the public mind to those feared from the nuclear industry. The European response to this event has taken the UK legislation, established after Flixborough, a stage further as a result of directives agreed in Brussels. It also stimulated the general public, already concerned about a whole range of perceived industrial impacts upon the environment. These concerns had given rise to and been further stimulated by well-organized pressure groups, whose campaigns on such issues as nuclear waste, lead in petrol and acid rain, have shown that large numbers of people may be politically motivated by their concern for environmental issues, notwithstanding the findings and merits of academic investigations and irrespective of the necessity for wealth creation and jobs. This has even led to the formation of new political parties across Europe with a measurable influence on the balance of political power.

Such public concerns are not confined solely to manufacturing activities; they extend to industrial end use, storage and waste disposal. Some of the worst disasters have taken place in these areas, with many fatalities from food poisoning in Iran, edible oil contamination in Spain and wine adulteration in Italy all making headlines in the international press. Many lesser incidents have featured in the national news, involving traders and hauliers transporting and storing hazardous materials under substandard conditions. A grim reminder of the potential harm from such malpractice was the death of over two hundred holidaymakers on a Spanish campsite in a single incident in 1978.

It is of course widely appreciated that these industrially based disasters are unlikely to cause such heavy tolls as are frequently reported from natural calamities such as earthquakes, volcanic eruptions and floods. But industrially based events are held to be preventable, and provide further support for the anti-industry cultural streak which characterizes an element of middle-class attitudes in the UK. It may also be acknowledged with the passage of time that some disasters are part of the price which has to be paid for technological progress and, while to some extent unforeseeable, nevertheless provide an invaluable base for future development. Thus society is now prepared to take a philosophical view over such disasters as the Tay Bridge collapse in 1879, taking with it a whole train and its complement. A generation or so later a similar view is taken of the disasters with the Comet jet aircraft. These early design errors did not stop the building of bridges (or even subsequent bridge collapses such as at Quebec or Tacomah), or of jet aircraft. Our society has come too far down the road of industrially based technological support systems to turn back, and probably cannot do so even if it wished. But at the present time it is hard to take a long view over such disasters as have recently occurred in a suburb of Mexico City, or at Bhopal in India, and there is an abiding apprehension about nuclear power. This fear of nuclear radiation from commercial power stations has developed in spite of much scientific evidence, in spite of an excellent safety record (even taking full account of Chernobyl), and in spite of its undoubted commercial and economic value.

The Mexican and Indian disasters tend to reinforce an impression given by international statistics that such major industrial causes of fatalities are more likely in the developing world than in the longer established industrial environments of Europe. Whether or not this impression is more apparent than real, there is an undoubted difference in the scenario which has to be constructed when assessing the risks and the benefits from advanced technological development in the developing and developed world.

Estimating the risks from major industrial hazards, and then assessing them in the light of their associated benefits and the costs of risk reduction or mitigation, involves many complex issues. Judgements made from partial assessments are likely to be much influenced by the position of the assessor. Major hazards invoke strongly emotive factors which may be influenced by presentations from investigative journalists: these may heighten legitimate public interest but they often confuse the real but difficult issues. Assessment of a major industrial hazard will never be a suitable bedtime story for children; neither for that matter is it likely to be adequately treated in a late-night news flash. On the other hand the report from the two-year-long public inquiry into the proposed Sizewell 'B' PWR (pressurized water reactor) station[5] seems likely to prove indigestible to the general public – even the daily proceedings proved so tedious as to be unreportable by the popular press.

To try to bridge the gap between these extremes would be foolish, and the aim of this book is to provide an introductory appreciation of present understandings of the key elements which need to be taken into account when attempting a risk assessment of a major industrial hazard. The book is written primarily for practitioners in industry and in the planning and regulatory authorities. Inevitably the treatment is mathematical in some areas, but the wide scope of the book has inhibited exhaustive descriptions and explanations. These may be found in the references provided at the end of each chapter. It is also hoped that the book will be intelligible to a wider readership. Participation by an informed public in matters of risk assessment will often be necessary to secure effective and enduring outcomes in cases where hard choices need to be made.

It is sometimes said that committed experts should refrain from attempting an overall evaluation of their subject, but if the limitations and significance of risk estimates are to be understood and explained, familiarity with the data obtained from practical experience is needed. Chapters 1 to 4 cover general questions of definition, classification, regulation and management. Chapters 5 to 9 explain how estimates are made of release quantity and likelihood, dispersion, ignition and explosion, damage and its distribution amongst a population. The concluding chapters deal with transport, mitigating hazards and with cost estimates.

Some of the matters, notably those relating to the design and construction of equipment, management, and the operation of processes, result from a lifetime of experience in industry. But some of the more specific aspects, notably those relating to toxic effects, blast damage and mathematical modelling, result from a recent period as a Research Fellow in the Department of Chemical Engineering at Loughborough University. Many ideas were formulated there as a result of conversations with close colleagues Trevor Kletz, Judith Petts and Frank Lees, who have each written authoritatively on the subject. However, it must be said that responsibility for the opinions expressed in this book belongs to the author alone.

References

1 Harvey, B. N. (Chairman), *Third Report of the Advisory Committee on Major Hazards*, London: HMSO, 1984.
2 Health and Safety Executive, *Guide to the Control of Major Accident Hazards Regulations*, London: HMSO, 1984.
3 Petts, J. I., *The Canvey Inquiries*, Working Paper MHC/85/1, Loughborough: Department of Chemical Engineering LUT, 1985.

4 *Rijnmond Report*, Dordrecht, Netherlands: Reidel, 1982.
5 Layfield, F., Sir, 'Sizewell B Public Inquiry: Report on Application by the CEGB for consent for the construction of a PWR', 8 volumes, Department of Energy, London: HMSO, 1987.

1 Risk analyses and risk perceptions

The most readily understood form of risk is the chance of death to an individual permanently located at a specific place. This simple definition of the 'individual risk' frequently provides the starting point for detailed numerical risk analysis. In the detailed analysis it is usually necessary to make distinctions between instant death and delayed death, and it may be required to take account of injuries and fates 'worse than death'. Since instant death is readily identified it is the most commonly found form at the start of a numerical risk analysis when undertaking a preliminary survey. The subsequent considerations may be factored from the figures for instant death, depending on the nature of the injuries and the state of our knowledge. Thus, for example, much data is available on the ratio of deaths to injuries arising from the blast and fire damage of air raids in World War 2,[1] but there is some uncertainty about the ratio of instant death to delayed death and injury arising from exposure to some toxic substances or to nuclear radiation. Usually the 'individual' is representative of an average fit and healthy person, and in detailed analyses it may be necessary to distinguish between various categories of 'individual', as well as take into account the actual time spent in the place and other factors.

Individual risk

The calculation of individual risk involves working through the long chain of events leading to the release and dispersion of a hazardous material to obtain the chance of death occurring to an individual permanently located at the defined point in space away from the source of the release. The final element in the long chain is the damage relationship which relates the chances of damage (death or injury in the case of individuals) to some dependent variable of the release such as concentration in the case of a toxic release or overpressure in the case of an explosion. As will be explained in Chapter 8 the damage relationship usually involves more than one dependent variable.

For a set of events the chances are simply added together to give a total frequency. Such a set may comprise a series of releases from the same source, of differing size and frequency, it may arise from a number of different sources or be a combination of the two.

Individual risks may be presented as a set of risk contours. These are lines joining points of equal individual risk around the source, the set usually showing a difference of a factor of ten between successive contours.

The concept of societal risk provides a measure of the chances of a number of people being affected by a single event or set of events. The chances reduce as the number of people increases and societal risks are often presented as *f/n* curves. Such curves are drawn as a graph where the abscissae are the numbers of casualties in an event (n) and the ordinates are the frequencies of the event (f). A cumulative frequency is often employed to present the frequency at which a population exceeding the given number is damaged.

Risk contours are useful in a preliminary study since they:

1 provide a direct means of mapping areas of risk;
2 permit a direct comparison with actuarial data and other risk criteria;
3 provide readily understandable criteria for assessing the effects of possible plant or system modifications, for example:
   (a) a reduction of maximum release quantity;
   (b) increasing the separation distances;
4 do not require any information on the distribution of the affected population, either in time or space.

But they cannot by themselves provide any indication of the number of people at risk.

For the purposes of emergency planning it is clearly necessary to provide an estimate of the number of people likely to be affected, and any communication to the general public is unlikely to be well received unless such an estimate is given. Although individual risk contours may well serve to distinguish areas of risk in a preliminary assessment, societal risk calculations will often be demanded if the assessment is to be seen as complete.

Such estimates also need to be given in the context of their likelihood. That the general public perceives events which affect larger numbers but happen infrequently with more alarm than those of higher frequency involving smaller numbers is well known. This may hold true even though the accumulated total of the latter may far exceed that of the former, and the location of the hazard source be exceedingly remote. A case in point is the comparison of the public anxieties in the United Kingdom between the single incident at Bhopal in India and road traffic accidents. Public anxiety following Bhopal was sufficient to stop the building of an unrelated chemical plant in Scotland, but it exerts little pressure upon traffic volumes. On the other hand the frequency of the worst case scenario involving large numbers of casualties may be seen as so low as to be negligible. This was the position perceived by the local authorities[2] at the public inquiry into the proposed building of another nuclear power station at Sizewell in Suffolk, but whether this would still be their view following the Chernobyl incident is a matter of opinion. Unfortunately the public's perception of frequency and magnitude of possible hazards bears little resemblance to predictions of expert risk assessments or even past case histories. This difference in perspective may be enhanced by the activities of the media and of pressure groups which thrive on confrontation and raising the level of public anxieties.

Whilst, therefore, the essence of societal risk is simply to provide an estimate of the number of people affected by a single event, there are particular difficulties in the presentation when more than one event is possible. Thus for some purposes it may be inappropriate to combine or accumulate the numbers of people affected from various sizes of release since this implies an acceptance of a linear dependence of total harm upon the number of casualties, which can be contrary to the public's perception of risk assessment. On the other hand the practice of accumulating industrial injuries and deaths on an annual basis for purposes of comparison and safety improvement is well understood by all concerned.

Numbers such as 10^{-7} may mean nothing to ordinary members of the public and even technologists can find it hard to form a mental picture. Trevor Kletz has suggested a number of ways in which risks can be explained, as for example listing life expectancies assuming we would all live for ever were it not for a particular risk:

1 if we were to smoke 40 cigarettes a day – 100 years;
2 a steel erector working 2000 hours a year – 750 years;
3 drinking a bottle of wine a day – 1300 years;
4 driving a car for 10 hours a week – 3500 years;
5 being struck by lightning – 10 000 000 years.

These are of course average figures.

Acceptable risk

Almost all human activities involve some risk, and public policies have to balance possible extra risks against ascertainable benefits, in some kind of trade-off assessment. Such an assessment may be required to aid decisions involving expenditures on risk mitigation or even whether an activity be allowed at all. In this context the concept of a risk threshold often appears, known as 'acceptable risk'. It is based on the assumption that there exists a non-zero level of probability of occurrence of an accident below which the public is willing to accept the risk. Fischoff *et al.*[3] have proposed the following definition for the concept: 'The acceptable level is the level which is "good enough", where the advantages of increased safety are not worth the extra costs of reducing risk.'

To help form an opinion about the level of risk which might be 'acceptable' to the general public, reference is sometimes made to actuarial tables giving accidental deaths attributable to occupational or other conditions. Some examples of actuarial data appear in Table 1.1.

Table 1.1
Actuarial data

Overall chance of death	
for man aged 30	1×10^{-3} per annum
for man aged 60	1×10^{-2} per annum
Extra chance of death	
in clothing manufacture	5×10^{-6} per annum
in chemical industry	8.5×10^{-5} per annum
in agriculture	1.1×10^{-4} per annum

Table 1.2
Risk reports on extra chances of death

Cremer and Warner's Rijnmond Report	
Deaths amongst employees at UKF ammonia plant	3×10^{-5} per annum
Deaths amongst the surrounding population	3×10^{-7} per annum
Royal Society's study group	
A figure for 'acceptable' risk	1×10^{-5} per annum

By way of contrast the data in Table 1.2 gives the results from a well-known risk assessment carried out for a public authority in the Netherlands,[4] and the 'acceptable' level recommended by a study group sponsored by the Royal Society.[5]

Thus, the assessed risks to the workpeople at the UKF plant are significantly lower than might be expected based on the data for the chemical process industry as a whole, while the assessed risk to the surrounding population is very much lower than the 'acceptable' risk put forward by the Royal Society's study group. A set of risk contours centred on the plant shows that the residential boundary is so far away from the plant that the contour which crosses it represents a risk of death of only once in a hundred million years. Although the population density beyond this boundary is high the assessed *f/n* curve shows that the societal risk of killing a thousand or more people is equally remote.

Further comments about the risk assessments for this situation appear in a later chapter, but Figure 1.1 provides a transect through the contours to give a typical illustration of the variation of individual risk with the distance of the defined place from the hazard source.

At the time when the Rijnmond study was executed there was some doubt about the lethal toxicity of ammonia, and the public authority directed the consultants to adopt a very conservative figure. More recent work suggests that the figure adopted was too severe by a factor of three. Furthermore, the risks given in Table 1.2 were calculated on the basis of people being out of doors whereas it could be expected that a drifting toxic cloud would allow time for an emergency plan to be in operation and for people to be indoors. This would be expected to reduce the lethal toxicity by a further factor of three. Thus the risk assessment for the UKF plant gave no cause for public anxiety. Moreover, ammonia is a

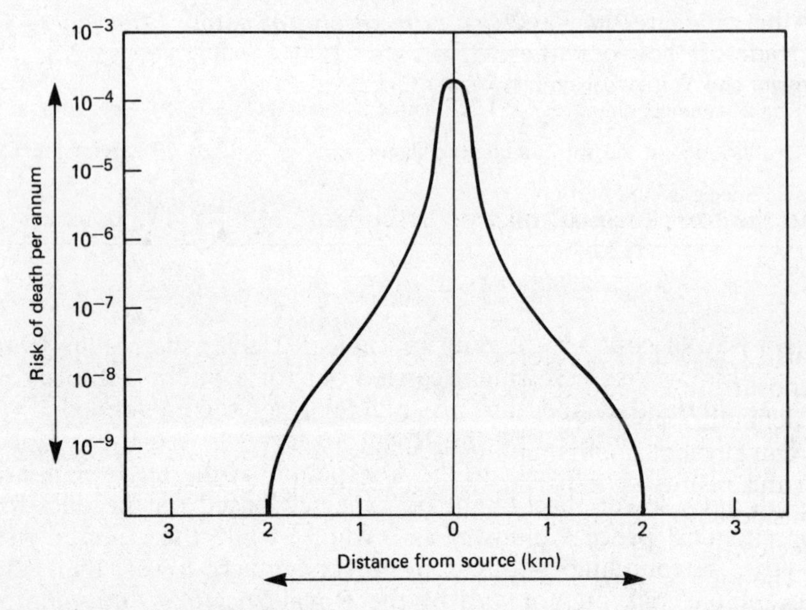

Figure 1.1 *A typical risk transect*

substance widely used in agriculture and industry, and its manufacture is in a mature phase following substantial expansion worldwide over the last fifty years. During this time a vast quantity of operational expertise has been amassed, which has tended to minimize the incidence and consequences of accidental releases. The public has, therefore, a large measure of confidence in the management of this type of plant and is well aware of the benefits it brings.

Thresholds and uncertainty

The Rijnmond risk assessment was, however, based on the likely consequences to average people and was weighted towards the use of average data based upon the predicted mortality to 50 per cent of an exposed population. An inherent weakness in all numerical risk assessments relates to the damage relationship for 'vulnerable' people, who make up a minority of an exposed population.

The injurious effect of the inhalation of a toxic gas is a function of the exposure time and the concentration. Within limits there is a trade-off between these factors so that for a given degree of injury the following relationship applies:

$$C^m T^n = \text{constant}$$

and the toxic load, L, may be described:

$$L = f(C,T)$$

thus LC_{50} and LL_{50} refer to the lethal concentration (at a specified exposure time) and to the lethal load for 50 per cent mortality of a defined population.

Our present understanding of inhalation toxicity is largely based on the results of experiments with animals. These are subject to considerable variation due to differences between specimens, species, and caging conditions.[6]

The interpretation of toxicity experiments has been discussed by Trevan, who drew attention to the characteristic shape of the line obtained when, for example, percentage fatality is plotted against concentration at constant exposure time.[7] This is illustrated in Figure 1.2, which gives results for ammonia at an exposure time of 30 minutes. There is a characteristically linear portion between the limits of 20 per cent and 80 per cent, with very non-linear effects outside these limits. The graphical method of Litchfield and Wilcoxon[8] which has been used frequently by toxicologists to determine LC_{50} values begins by fitting the best straight line to results obtained over the range 20 per cent to 80 per cent, and they recommend the use of log-probability paper which tends to provide a straight line over a wider range of values and which may facilitate the extrapolation of the results to lower percentage fatalities. Figure 1.3 provides such an example on log-probability paper using the same data as Figure 1.2.

In recent years purely numerical methods of predicting the effects of toxic release have become commonplace. These involve the use of a probit equation:

$$Y = k_1 + k_2 \log x$$

where Y is the probit and x is the toxic load.

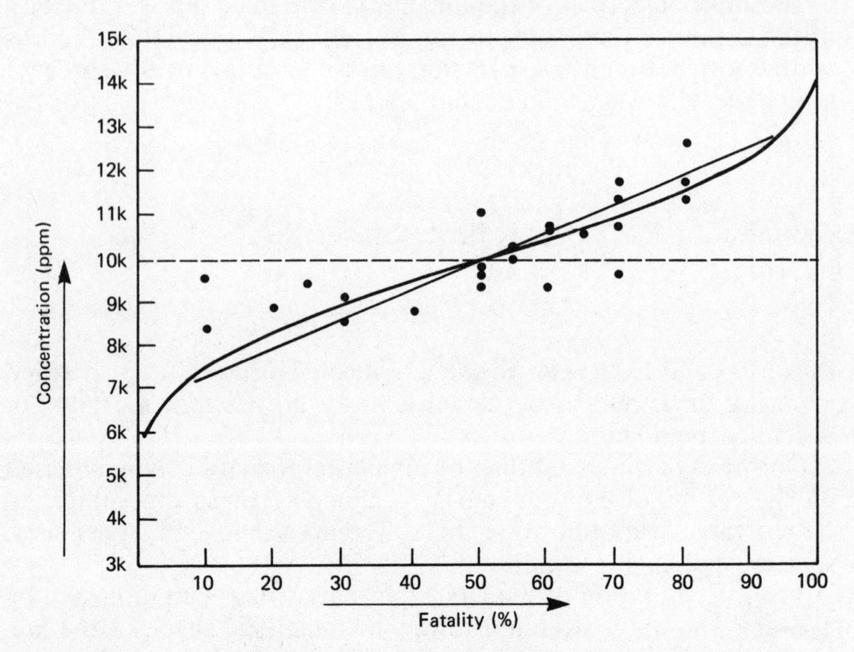

Figure 1.2 *Lethal toxicity of ammonia, linear plot*

A more detailed account of probit equations is given elsewhere.[9] Probit methods provide a very convenient way of extrapolating from data obtained over the range 20–80 per cent mortality, but it must be remembered that their use does not confer any greater precision than is inherent in the original data. Thus, there is obvious uncertainty in the slope of the line drawn through the experimental points obtained in the 20–80 per cent region and extrapolation of this uncertain line to levels of one per cent or less is open to large errors. Very large numbers of animals would have to be employed to obtain significant experimental data in the low percentage region.

Trevan also drew attention to the low level of confidence which may be given to the mean result if too few animals are used. It has been shown using Trevan's method that for small numbers of animals the confidence limits are wide even at the central 50 per cent mortality and that they become ever wider at the lower (and higher) mortalities. This is illustrated in Table 1.3, reproduced from a paper by Withers and Lees, giving ranges at the 95 per cent confidence level.[10]

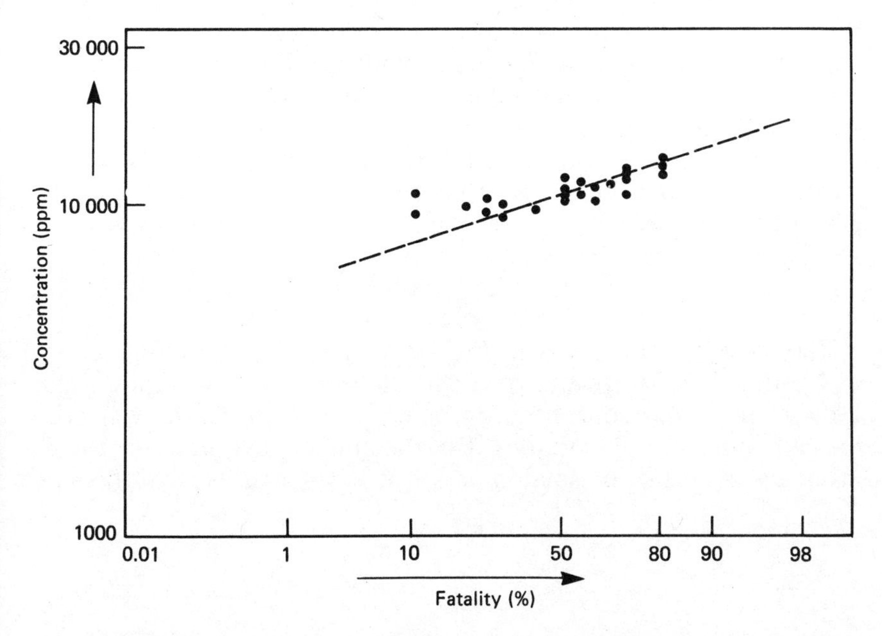

Figure 1.3 *Lethal toxicity of ammonia, log-probability plot*

Whilst it is clearly important that experiments be conducted with large numbers of animals, humanitarian and economic considerations usually provide a limiting constraint. As a result, even though considerable care is often taken to work within an

Table 1.3

Expected mortalities in groups of experimental animals, at 95 per cent confidence level

Mortality expected (%)	Number in set	Fatality range
50	10	2–8
	20	6–14
	30	10–20
	50	18–32
	100	40–60
25	30	3–12
	100	17–33
10	30	0–6
	100	4–16

agreed protocol using carefully selected specimens, there is an appreciable variation in published findings. Figure 1.4 illustrates this with further data on ammonia, taken from various authors, showing derived LC_{50} values over a range of exposure times.[11]

The most realistic line through this data display is based on a mean LC_{50} at 30 minutes of 1.15 per cent and a toxic load formula, which has the form:

$$C \times T^{\frac{1}{2}} = \text{constant}$$

The shape of the curve in Figure 1.2 and the extrapolated line of Figure 1.3 is suggestive of a threshold value of around 0.5 per cent in the lethal concentration below which no deaths will arise for an exposure of 30 minutes. But the animal experiments provide instances of death for small animals at lower figures than this and

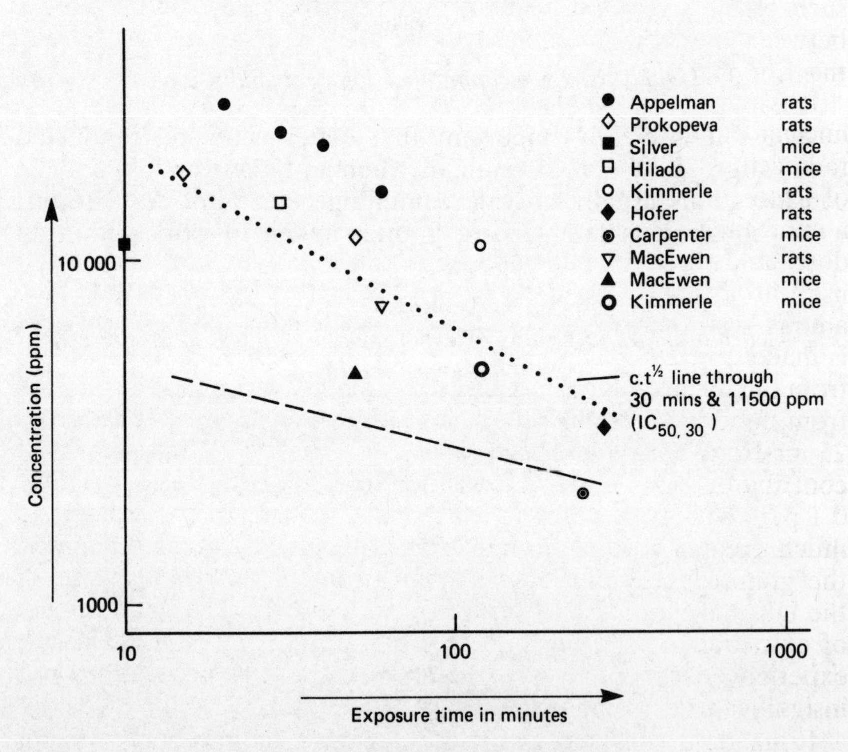

Figure 1.4 *Lethal toxicity of ammonia, LC_{50} values*

the uncertainty requires that such a threshold figure be treated with caution, particularly when the presence of vulnerable people is recognized. Animal experiments have demonstrated a large variation in response between apparently equally healthy members of the same species to toxic effects. One will have walked away apparently none the worse for an exposure, another will have died within hours. In the case of human beings, it is known that elderly people and the very young are particularly vulnerable to the toxic gases.

This example of ammonia has been used to demonstrate some principal difficulties in aspects of risk analysis which are common both to toxic gases and to the effects of blast and thermal and nuclear radiation. While this is a common problem to all damage relationships, discussion so far has been confined to the assessment of the risks of instant death. Consideration of the risks of delayed death or injury provides similar further difficulties in the assessment of particular hazards to small percentages of the population. In such cases it may be necessary to provide a substantial margin between the lethality or injury level used in the risk analysis and the LC_{50}.

It is particularly difficult to assess the effects of small doses of nuclear radiation. We live in a highly variable background level of radioactivity, which in the UK gives an average dose per person of about 0.215 rems or 215 millirems (mr) each year. (There are a number of units in common use for the measurement of radiation doses and these are explained in Chapter 8.) The rem is the term used in advice to the British government, and is in common use among scientists in the USA and Soviet Union. The annual dose is made up from around 80 mr in the air we breathe, about 40 mr from the material surroundings (for example, rocks and soil), 37 mr from food and drink, 30 mr from rays from outer space, and 25 mr from X-rays and medicine. The UK nuclear industry itself contributes only some 0.2 mr per annum, equivalent to some 0.1 per cent of the normal background.[12] There is, for example, a much greater dose to be had from a fortnight's holiday amongst the granitic rocks of Cornwall than from the normal activities of the UK nuclear industry over a ten-year period.[13] The small amount of extra radiation from the Chernobyl reactor disaster, so briefly experienced by an average member of the UK population, was insignificant by comparison.

From data collected about the Japanese atomic bomb victims and from animal experiments, estimates of the lethality of varying doses of nuclear radiation have been made. Thus, instant or early deaths result from doses of nuclear radiation of around 1000 rems (1 000 000 mr). The main agent of delayed death is the inducement of an extra risk of cancer or leukaemia over and above the 22 per cent average chance of dying from these causes (in the UK) rather than from some other cause. It is predicted that if 200 times the average annual dose (200 × 200 mr = 40 rems) is received by one person over a few days, the chance of dying of cancer (in the UK) before death from any other cause is increased from 22 to 22.5 per cent. A dose of this magnitude administered to each of 200 people is therefore likely to cause one extra death by cancer, but it is not possible to predict which one amongst the 200 would be affected.

For much smaller doses the consequences are obscured by a variety of indeterminate factors which make detailed analysis impossible. It is possible to extrapolate the above figures by assuming that a linear relationship applies and that on average the same single death will occur if one-thousandth of the dose is received by 200 000 people, but this makes no allowance for the biological defence mechanisms which operate at these lower levels in humans, particularly when the doses are received over much longer periods and included in the natural background radiation. It is in this context that the Director of the UK National Radiological Protection Board is reported as saying that the Chernobyl disaster could result in some 'tens' of extra cases of cancer emerging in the UK over the next fifty years amongst the many millions of cancer cases that would arise from natural causes. Even a thirty-three-year-long study of 40 000 UK Atomic Energy Authority employees has given inconclusive results. While it suggests that the mortality rates from cancer were actually lower than for the general population, the sample size provided by even this large group of workers was not big enough to provide unambiguous proof on a statistical basis.

To avoid the obvious pitfalls in trying to apply formulae in such circumstances, the International Commission on Radiological Protection has recommended limits which are not too far removed from natural background levels and which are far below the levels for which deleterious consequences have been observed. This is in line with the practice of most regulatory bodies concerned with safety matters, which are expected by the general public they serve to take a cautious and conservative view.

An interesting account of the way in which the UK Health and Safety Executive (HSE) arrives at its criteria for chlorine toxicity in the context of major hazards appeared in the monthly journal of the Institution of Chemical Engineers in 1985.[14] As has been explained previously for ammonia, there is considerable uncertainty about the toxicity relationship although there is enough data to arrive at appropriate levels of confidence and to obtain a consensus figure for an average LC_{50} for chlorine at 30 minutes' exposure time.

This consensus figure is in the range 300–400 parts per million, and from the available data the HSE derived a lower 95 per cent confidence limit of fifty-seven parts per million for an LC_{01} at 30 minutes' exposure.

This concentration is below the levels where lethality had been observed in animal experiments covering a variety of species. Extrapolation of data plots on log-probability paper of the type illustrated in Figure 1.3 gives a best estimate in the region of 200–400 parts per million for the LC_{01}.

Realism and conservatism

Because of the uncertainties in the data and the difficulties in allowing for the more vulnerable people, the HSE is bound to take a fairly cautious view and, in making its recommendations, to incline towards the use of values such as the lower level of a defined confidence limit. On the other hand, industry is anxious that the risks from potential major hazards should not be exaggerated, and presses for realistic assessments wherever possible. It may seek to avoid the presentation of the results of numerical analyses altogether where the confidence limits are widely spaced and the predicted frequency is based on a probabilistic assessment from 'generic' data.[15] In these situations industry often prefers to quote from its records of site-specific safety management to limit the overstatement of cases by the media and pressure groups.

A degree of caution is appropriate when dealing with a prospective major hazard. Nevertheless, in the first instance a risk assessment should aim to be realistic rather than conservative. However, information should be provided on the sensitivity of the result to uncertainties in the assumptions made. The estimate can then be viewed with an appropriate degree of conservatism. If, on the other hand, conservative estimates are introduced at every point in the risk assessment chain the final result may be so unrealistic as to be absurd.

How far a risk assessment should go depends upon the nature and level of the risks and the public perception. It seems unlikely, however, that a risk assessment presented to a public inquiry would be considered complete without estimates of individual and societal risks.

Differences in expert estimates both of the frequency and the consequences of accidents are inevitable, but may be made more understandable by a more explicit recognition of the different types of data which may be used. For example, when estimating the frequencies of possible failures use is often made of records of past failures of similar equipment. This use of 'generic' data is very necessary at the outset of a preliminary survey before detailed engineering data specific to the proposed site is available. Databanks of the 'generic' reliability of components provide a valuable input to the subsequent more detailed and site-specific studies of a full risk analysis. Predicting future failures from the statistical analysis of historical records may, however, give a conservative figure for new installations, because it may not make sufficient allowance for technological improvement based on the working experience. Particular care needs to be taken when using 'probabilistic' methods of risk analysis to forecast a very rare event. An alternative technique of making a predictive mathematical model based on scientific and engineering principles is usually preferred. Both can contribute to a decision but the former is liable to give a higher figure than the latter, perhaps by several orders of magnitude.

An exhaustive debate[16] on such issues took place at the Sizewell inquiry, where evidence on the results of a historical survey gave a probabilistic failure 'upper bound' rate in the region of two in a million reactor years. The CEGB objected strongly to the use of such an estimate as the basis for the top event failure rate, and advocated instead a figure derived from a deterministic mathematical model based on stress analysis. This gave a figure one hundred times lower than the probabilistic 'upper bound'. Independent evidence presented at the inquiry by local authorities and others supported the deterministic approach provided it was allied to strong management measures in quality assurance through all the procurement phases. This requires much engineering time and effort but is nevertheless necessary for installations which might otherwise constitute a major hazard.

Differences over consequence estimates may derive from the damage relationship employed or from the population distribution that has been assumed, but the differences may be less intractable when their nature is properly explained and the sensitivity of the final outcome to the uncertainty made clear. Moreover, there are signs that both in frequency and in consequence the differences are becoming less. The figures quoted for the frequency of pressure vessel failure show a converging trend while the range of hazard distances given by the main gas dispersion models has narrowed appreciably, as is demonstrated in a later chapter.

Societal risk

Whereas in the case of individual risks, the risk analysis is generally directed towards the twin ends of providing risk contours and drawing comparisons with actuarial data of 'acceptable' risk, the analyses devised for societal risk have a number of objectives in mind. These include:

1. tabulating societal deaths per annum for specific locations and from specific sources;
2. tabulating accumulated societal deaths per annum for all locations and all sources;
3. tabulating fatalities against release size, displaying the outcome as a graph on logarithmic co-ordinates;
4. displaying frequency/fatality magnitude graphs (*f/n* curves) to distinguish size from frequency: in this connection cumulative frequency may be used, that is the frequencies of all the events which exceed a given number of casualties may be added together;
5. repeating any of the above so as to show the differing effects of such consideration as:
   (a) daytime as distinct from night-time populations;
   (b) indoor as distinct from outdoor risks;
   (c) more vulnerable population elements as distinct from less.

As in the evaluation of individual risk, the damage to people is usually calculated in terms of instant or early death. As stated at the beginning of this chapter, there are a number of qualifications to this concept which may need to be taken into account, but it has the great merit of unambiguity and estimates of total casualties can often be derived from the predicted fatalities. There are of course several classes of risk where the public may perceive a fate worse than death, and it is often remarked that no assessment of societal risk is meaningful unless benefits are considered as well. The impact of such considerations upon the presentation of societal risk may be considerable.

A structure for risk assessment

A phased approach is preferable when undertaking a risk assessment, with separate considerations being given to on- and off-site casualties. Emergency plans need to be made and rehearsed for major hazards, and in addition to the possible number of fatalities an estimate of the number of injuries will be required. This may be derived from the estimate of fatalities. Descriptions of the ways in which the detailed calculations of individual risk and of societal risk are made appear in later chapters. Individual risk contours provide a helpful initial screen by which a preliminary assessment of a perceived problem can be formulated.

A full assessment involves many aspects and needs to be viewed from a number of positions by people with different backgrounds and perspectives. There is a need for an agreed structure for this purpose to provide some common ground. A four-part structure has been proposed as follows:[17]

1 a broad definition of the processes and systems involving hazardous materials on site, giving quantities and storage modes;
2 a statement of possible releases of hazardous material with an estimate of their likelihood and magnitude;
3 a summary of emergency scenarios to give an indication of possible chains of events from release to property damage and injury to people;
4 an assessment of these estimates with comment on accuracy and criteria against which societal risks may be judged.

Societal risk is probably the most important issue in the public mind, related to its fear of catastrophic events. Criteria for the evaluation of societal risk are still under debate but the Groningen *f/n* lines are very widely recognized and often applied. These lines were proposed by the Provincial Waterstaat, Groningen, Netherlands, and are reproduced in Figure 1.5. They separate three categories of risk: 'acceptable', 'needing further assessment' and 'unacceptable'. As Hagon has remarked,[18] it may be that current doubts about the value of risk assessment are illustrated by the differing totals of implied fatalities beneath the two lines. The 'acceptable' figure is around 1.5 fatalities in a million years; the 'unacceptable' figure is around 1.5 in a hundred years.

The Groningen 'unacceptable' line closely follows the UKF ammonia *f/n* line of the Rijnmond Report referred to earlier and may be thought too onerous, especially when considered in the light of the second Canvey Report whose findings are represented by the *f/n* data also illustrated in Figure 1.5. But the underlying concept of the three categories must be sound. Mathematical modelling of typical potential hazards in densely populated areas suggests a very rapid change in the levels of societal risk over comparatively short distances from the hazard source so that the separation of the two lines may not be unreasonable.

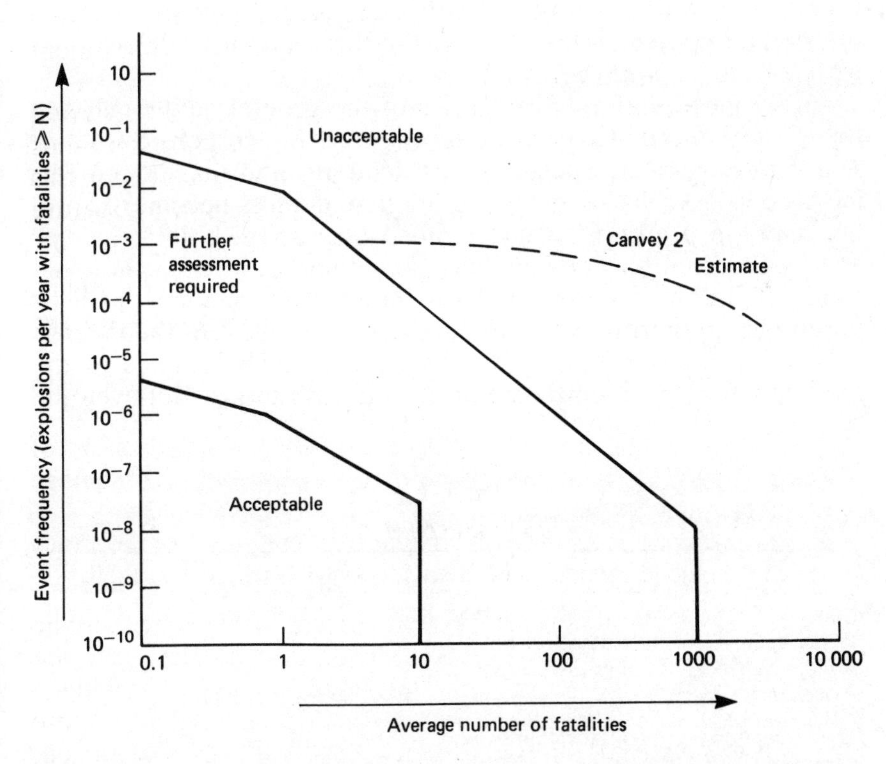

Figure 1.5 *Risk criteria provided by the Provincial Waterstaat, Groningen*

Evaluating the results of the risk assessment using a set of risk criteria involves value judgements and, inevitably therefore, disagreements. It is helpful to put the results into a context to clarify issues. The four-part structure given above suggests that the results be compared with the corresponding criteria for individual and societal risk. The debate on risk criteria will continue but it does not now seem hopeless to envisage the growth of an understanding of the use of risk criteria as an aid to decision-making.

Public perception of risk

When the general public is asked to quantify its preferences, a quite different perspective from that of technical experts may emerge as Table 1.4 makes clear.[19] It compares the disease and accident estimates of Inhaber relating to the production of electricity for oil, coal and nuclear power with the general public's perception of estimated potential fatalities for the three systems following the work of Slovic *et al.*

Furthermore, differing categories of the general public may see things very differently from each other. At the corporate level of senior management, Spetzler[20] has demonstrated that as projects increase in risk, that is as the size of gain and loss becomes larger, managers in a group become more conservative. Utility theory[21] and behavioural science may have something to offer in the long-term resolution of these problems but, for the moment, at least the French authorities seem to have overcome most of their public aversion to nuclear energy by providing tangible benefits to the local communities[22] which can be seen in cash terms. But even the

Table 1.4
Differing perspectives

	Coal	Oil	Nuclear
Expert estimate of man-days lost per megawatt year through disease and accidents	2000	1900	1.5
Lay perception of potential fatalities per annum	17×10^3	17×10^3	27×10^6

value of money to individuals as compared to its face value can vary non-linearly and in absolute terms from one to another, so that it is not surprising to find large differences between people in the assessment of risks and benefits. An attitude survey carried out for the HSE in 1982[23] suggested that about one-third of the general public in the UK does not think that chemical or nuclear plants provide any risk to it, but of the two-thirds who do, the great majority felt that they would have to be a substantial distance away (perhaps 50 miles or more) to be free from worry. It also seems that the nearer someone lived to a major industrial site the less likely would he consider there to be a risk to the public. Furthermore, the readers of 'quality' newspapers and members of the higher income families seem to feel the highest likelihood of a serious mishap. Perhaps these responses are in line with the vigorous opposition to the proposed petrochemical plant at Canvey (where there is a significant 'safety' distance of waste land between the plant and the residential boundary), and the apparent absence of complaints about the major plants in Cleveland and Cheshire in the north of England, where the housing is often close to the works perimeter.

The often expressed fears about long-term hazards and the increasing perceptions of major catastrophes are not reflected in official records. Typical figures appear in Table 1.5.

Over this period Kletz and Turner[24] have counted the number of incidents occurring internationally within the oil and chemicals industries involving five or more fatalities. They found a roughly constant rate of seven per year averaging twenty-one deaths per

Table 1.5
Mortality data

Category	Ten-year total
All UK accidental deaths	191 942
All UK road accident deaths	78 968
All UK manufacturing industry deaths	2 365
All UK petrochemicals industry deaths	260
Number of petrochemicals industry incidents with five or more deaths	1
Number of deaths to the general public from petrochemicals industry incidents	0

incident, estimating that the industry had more than doubled in size during this time.

Kletz has also provided some information on longer-term deaths.[25] For 1975–78 the annual average of deaths from industrial accidents was 694 and that from prescribed industrial disease was 925. However, about 800 of the 925 were due to pneumoconiosis, asbestosis and byssinosis, diseases which are restricted to a few occupations and industries. If a comparison is made between days lost, the number due to industrial disease is only four per cent of those due to industrial accidents in 1976/77.

Another commonly expressed fear is the likely frequency of a predicted hazard. Even when the risk analysis suggests a frequency so low as to be negligible, the fear is that the event might happen tomorrow. But as Lees has pointed out,[26] before a low frequency major event occurs there is likely to be a series of lesser accidents and alarms, which happen at a relatively higher frequency. There is a likely sequence to such events which can provide the basis for a 'hazard warning structure'. Effective training and organization of workers in the plant will take note of these precursors to a catastrophe and initiate the appropriate remedial action. Where such effective management is missing, and warnings are not heeded, a catastrophe such as happened at Bhopal may arise anywhere. If the warning events are not happening, and the plant has established a reasonable period of steady-state operation, there is good assurance that the major accident will not occur either.

Among the assumptions made in any risk assessment are those about the quality of management. It may be argued in consequence that the assessment should be based on a non-conservative estimate. This ought not to be an argument for making the assessment itself more conservative but may be an argument for making a proper allowance for the management factor.

The public probably recognizes that an evaluation of risks involves choosing between things which are not strictly comparable under conditions of uncertainty. This is a task which industrial management faces every day of the week, as indeed do members of the public in their daily life. Technical experts should bear in mind, however, that particular risks produce particular fears which may not be open to direct comparison, and there remains much room for greater understanding of the risk perception of particular hazards.

Risk assessment is not necessarily the best form of reassurance which can be given to the public. More effective is a clear, first-hand perception of a strong safety policy executed under the control of effective and experienced management. However, once a plant has fallen into disrepute in the eyes of the public, not even a hazard warning structure will help.

It may be that a section of the public is not prepared to play the experts' game and rejects the whole concept of risk assessment, saying simply that any such risks are too great. However, at the Mossmoran public inquiry[27] in particular, there were signs that if the assessment process could be trusted and if the estimates were sufficiently low, the level of concern would fall away. Installations need to be seen to be safe in practice as well as in theory, however, and it is unlikely that any written statement can give the reassurance that comes from a satisfactory relationship developed in close proximity over many years.

References

1 Withers, R. M. J. and Lees, F. P., 'The lethal effects of a condensed phase explosion in a built-up area', LUT Report MHC/86/3, *Journal of Hazardous Materials*, 14, 1986 (under review).

2 Sizewell 'B' Power Station Public Inquiry, Proceedings day 243, and LPA 4 Add. 2, 1984.

3 Fischoff, B., Slovic, P., Lichtenstein, S. and Combs, B., 'How safe is safe enough?', *Policy Sciences*, 9, p. 127, 1978.

4 *Rijnmond Report*, Dordrecht, Netherlands: Reidel, 1982.

5 The Royal Society, *Risk Assessment*, Study Group Report, p. 180, 1983.

6 Withers, R. M. J., *First Report of MHAP Toxicity Panel*, Proceedings of International Chlorine Symposium, London 1985, Chichester: Ellis Horwood, 1986.

7 Trevan, J. W., 'The error of determination of toxicity', *Proceedings of the Royal Society*, B101, p. 483, 1927.

8 Litchfield, J. T. and Wilcoxon, F., 'A simplified method of evaluating dose–effect experiments', *Journal of Pharmacology and Experimental Therapeutics*, 96, p. 99, 1949.

9 Finney, D. J., *Probit Analysis*, Cambridge, Cambridge University Press, 1971.

10 Withers, R. M. J. and Lees, F. P., 'The assessment of major hazards: the lethal toxicity of chlorine: Part 1, Review of information on toxicity', *Journal of Hazardous Materials*, 12, p. 231, 1985.

11 Withers, R. M. J., 'Second Report of MHAP Toxicity Panel',

I. Chem. E. NW Branch Symposium Papers No. 1, 6.1, 1986.
12 Wright, P. and Prentice, T., 'Inside Chernobyl's cloud of confusion', *The Times*, 7 May 1986, p. 11.
13 Wrixon, A. D., 'Human exposure to radon decay products', *Atom*, 352, p. 2, 1986.
14 Davies, P. and Hymes, I., 'Chlorine toxicity criteria for hazard assessment', *The Chemical Engineer*, 415, p. 30, 1985.
15 *CONCAWE Report No. 10/82*, Fourth International Symposium on Loss Prevention, *I. Chem. E. Symposium Series 80*, 1, B1, 1983.
16 Op. cit. note 2, Proceedings days 216, 230 and 243.
17 Petts, J., Withers, R. M. J. and Lees, F., 'Expert evidence at inquiries into major hazards', *Project Appraisal*, 1, p. 3, 1986.
18 Hagon, D. O., 'Use of frequency-consequence curves to define broad criteria for major hazards', *I. Chem. E. Chem. Eng. Res. Des.*, 62, p. 381, 1984.
19 Schwing, R. C., *Trade-offs*, Societal Risk Assessment (Symposium at General Motors), p. 124, New York, Plenum Press, 1980.
20 Spetzler, C. S., 'The development of a corporate risk policy for capital investment decisions', *Institute of Electronic and Electrical Engineers (US) Trans-System Science & Cybernetics*, SSC4, p. 279, 1968.
21 Tribus, M., *Rational Descriptions, Decisions and Designs*, p. 344, Oxford: Pergamon, 1969.
22 Collier, J. G., 'Nuclear energy', *I.E.E. Electronics & Power*, 30–8, p. 615, 1984.
23 Health and Safety Executive, *Public attitudes towards industrial, work-related and other risks*, Social and Community Planning Research, 1982.
24 Kletz, T. A. and Turner, E., *Is the number of serious accidents increasing?* ICI Safety Note 79/2B, London: Chem. Ind. Assn., 1979.
25 Kletz, T. A., *Now or later? A numerical comparison of short and long-term hazards*, Fourth-International Symposium on Loss Prevention, *I. Chem. E. Symposium Series 80*, 1, A1, 1983.
26 Lees, F. P., 'The hazard warning structure of major hazards', *Transactions Institute of Chemical Engineers*, 60, p. 221, 1982.
27 Scottish Development Department, *Report of the Public Inquiry into the Shell/Esso Mossmoran/Braefoot Bay proposals*, 1978.

2 A classification of major industrial risks with some case histories

Later chapters will provide a more detailed understanding of the numerical elements of risk analyses. To help appreciate the context of these elements, various types of major hazard will now be discussed and classified. Chapters 3 and 4 provide background information on the legal position and management roles, and Chapter 11 discusses ways in which major hazards can be mitigated or prevented.

Major hazards are classified to distinguish three differing aspects. First, there is the nature of the hazard with respect to its process origin; second, the risks as between members of the public and workers on site; and, third, a distinction needs to be made between those major hazards which happen at the point of manufacture and those which may occur in the distribution and use or in waste disposal.

Explosive, fire, toxic and nuclear hazards

A number of comparative studies and lists have been published to show the relative magnitudes of the four categories of damage associated with fire, explosion, toxic release and nuclear accident. These suggest that, in so far as fatalities are concerned, fire is by far the most important. Blast fatalities are often associated with fire, and at about eight per cent of the total are about the same as toxic release fatalities although the latter presents a significantly greater hazard if injuries are taken into account. As has already been remarked, nuclear casualties have been negligible in comparison.

Blast incidents

The largest known explosion to occur in the UK happened at a munitions store near Burton-upon-Trent in 1944, when over 2000 tonnes of TNT blew up. Adjacent to the source, seven people were killed in a farm, which was completely obliterated. A further twenty-one fatalities occurred at distances up to 6.3 miles away, in or near buildings which were all practically demolished, and within this radius there was extensive damage to property generally and a covering layer of debris. Of course, many more fatalities would have occurred with a comparable TNT explosion in a built-up area. At a Silvertown munitions factory in 1917, for example, a 53-tonne explosion of TNT killed sixty-nine people, mostly within 250 metres of the source. Beyond 650 metres the damage to houses was described as slight.[1]

More typical of the type of accident that happens in the chemicals industry is the explosion which took place at 4.20 a.m. on 20 January 1968 at the Shell refinery in Pernis, Rijnmond, Netherlands. It followed a maloperation of a large 'slops' tank and the consequential formation of a hydrocarbon vapour cloud. The amount of material in the cloud has been estimated at about 50 tonnes, and the TNT equivalent estimated from the resulting devastation was 20 tonnes. Two people were killed in the industrial area around the refinery, but none in the residential districts. Altogether eighty-five were injured, mainly from flying glass. As is often the case with blast, the damage to property showed a number of anomalies. Some damage occurred as far away as Maasluis (8 km), whereas other much closer property at Hoogvliet and Pernis escaped.[2]

Fires

An examination of the large financial losses incurred by the chemicals industry from major disasters suggests that some 30 per cent of the loss was caused by fire, 68 per cent by explosion and 2 per cent from other causes. However, many of the explosions were followed by fire which made a major contribution to the explosion loss. Fire causes death in two main ways, asphyxiation or radiation burns. The former is most likely to occur when people are trapped by fire in confined spaces, the latter in the open.

The conclusion to be drawn from studying the findings of case histories is that far fewer deaths are caused by blast than by fire; deaths from primary blast alone are very rare. This, for example, was the conclusion of Zuckerman's work during World War 2 on air raid casualties.[3] A 1968 paper by Settles to the New York Academy of Science discussed the results from 81 chemical and gas accidents.[4] Forty-four involved fire and explosion, 23 fire only, and 14 a detonation reaction. Of the 78 fatalities, 19 died exclusively from fire, 58 died from a combination of fire and fragments, and only one person died from the blast wave (due to whole body translation). The TNO (Netherlands Organization for Applied Research) report on the San Juan Ixhuatepec LPG incident on 19 November 1984 in Mexico City[5] concludes that there was only minor damage from the blast wave, but that the greatest damage was probably caused by fire and explosions of the gas that accumulated inside houses.

Flixborough

The most noteworthy incident involving fire and explosion to occur in the UK in recent times was at the Nypro plant at Flixborough,[6] in June 1974, when a temporary pipe collapsed releasing about 50 tonnes of cyclohexane causing explosion and fire which destroyed the whole plant. There was extensive blast damage to houses nearby, but all the 28 dead were employees, of whom 72 were on site at the time. Altogether 108 people were injured and the damage was estimated at £100 million. Although this incident did not cause any deaths among the general public, it received much publicity, aroused widespread concern and had much influence on future official activity and legislation. The resulting inquiry revealed many inadequacies in the quality of the management and engineering procedures in this semi-nationalized concern.

Feyzin

An incident in France demonstrated a typical sequence of events leading to catastrophe. On 4 January 1966, at Feyzin refinery, an operator made some mistakes in opening and closing valves, which resulted in the uncontrolled escape of propane from a large storage sphere.[7] This froze the exit valve so that it could not be closed.

The alarm was raised and steps taken to stop traffic on a nearby motorway.

The high-density, very cold vapour cloud spread towards the motorway in a layer about one metre deep. It is believed that a car, about 160 metres away, had been left with its engine running and flames were seen to flash back from the car to the sphere in a series of jumps. The sphere was soon enveloped in a fierce fire. The pressure relief valve on the sphere opened and the additional escaping vapour also ignited. The sphere was one of eight similar storage spheres placed together. The fire brigade concentrated their efforts on attempting to keep the remaining spheres cool. After an hour and a half the sphere ruptured, killing the men nearby, toppling the sphere next to it and sending a wave of liquid propane to multiply the destruction. Altogether five of the spheres were destroyed, there were 21 deaths and 81 injuries. As is usual in such cases it was the firemen and the local workpeople who suffered, although some blast damage was caused to houses in the nearby village some 500 metres away.

San Juan Ixhuatepec

The worst accident involving fire and explosion in recent times occurred in Mexico City, where the destruction extended far beyond the confines of the industrial site, killing hundreds of the general public. The distance to the far border of the area which was most affected and where most of the houses were destroyed was about 300 metres. Within this area approximately 500 people were killed and altogether 7000 were injured. The majority of those killed were still asleep when they were surprised by fire and did not survive because of direct flame contact, the heat, the fumes, and the lack of oxygen. The radius of 300 metres agrees reasonably well with the TNO-calculated fireball radius that may have occurred, and the percentage mortality of the people living within this seems to have been as high as that for the German cities subjected to fire storms in the air raids by the RAF in World War 2.[8]

Figure 2.1 is taken from a computer display of a street plan of the San Juan Ixhuatepec area showing the damaged area where fatalities occurred. Also shown are risk contours based on an assumption of standard conditions for the quantities given in the TNO report. The damaged area is outside the 1/100 000 risk contour considered acceptable by the Royal Society's study group.

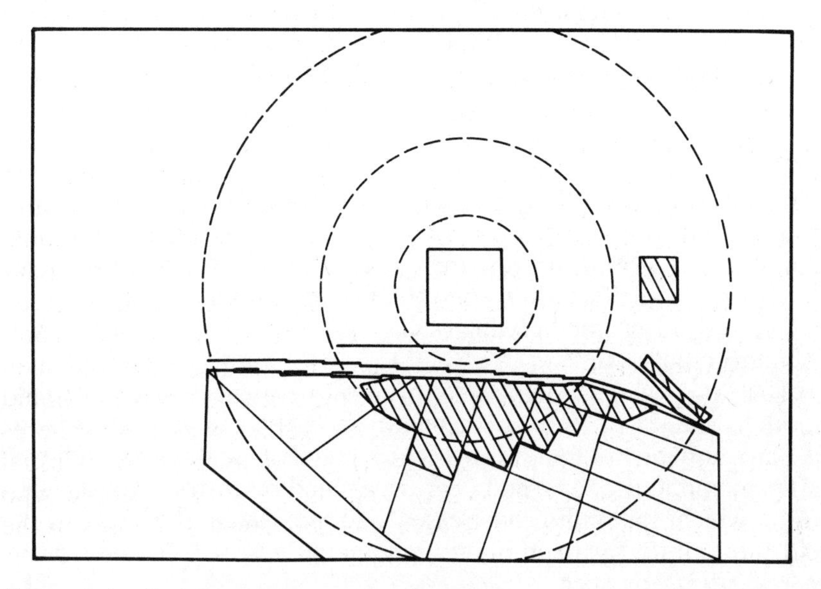

Figure 2.1 *San Juan Ixhuatepec street plan*

Nevertheless, many would consider that the housing and storage area were much too close together.

Toxic hazards

The previous chapter highlighted the potential hazard of a particular ammonia plant in the Netherlands. Ammonia is a common substance widely used in agriculture and industry. It is produced in a great number of locations throughout the world and may be stored in very large tanks with a capacity of perhaps 25 000 tonnes or more.

It is transported by all methods, in bulk as well as in cylinders and drums, with many end uses outside the confines of the manufacturing plant. Accidental releases causing fatalities have taken place during production, storage, transport and end use. Several sources provide records of significant accidents over the last thirty years, and these show that, worldwide, some 225 people have died, over one-third of these in North America. From 1971–77 there were 61 transport incidents causing injury or death in the USA alone. The largest number in any one incident was 21 in

Cartagena, Colombia in 1977. In the UK the largest incident was in London in 1941, when 16 people died, since when a further four people have been killed, the last in York in 1977.

Potchefstroom

An incident in South Africa typifies the kind of problem which caused many of the early failures, but has now been overcome. Ammonia is stored in pressure vessels which must be constructed to a special specification to be reliable. In particular, they must be fabricated from metal which will not suffer from fatigue and develop cracks.

The tanks at the Potchefstroom fertilizer plant, though small by modern standards, were built in 1967 and the dished ends were in a hard and brittle condition.[9] They had not been stress relieved after manufacture, and had been weakened by further weld repairs from which the induced stresses had not been removed. They did not conform to the current metallurgical and fabrication specification for such vessels.

On 13 July 1973 one of these tanks failed abruptly, about 25 per cent of one end coming apart. An estimated 30 tonnes escaped from the tank, forming a gas cloud about 150 metres in diameter and nearly 20 metres deep. One employee, 45 metres from the tank, was killed instantly by the blast; eight were killed by gas within 100 metres of the tank whilst attempting to escape; three more escaped but died after a few days. The cloud of gas reached some 300 metres in width and headed towards a nearby township; outside the plant boundary fence four people died immediately and two others several days later. In addition to the 18 deaths, 65 people required medical treatment in hospital. The cloud eventually reached a visible limit of 450 metres from the tank before it dispersed. The verdict at the inquest was that no one, by act or negligence, was responsible for the fatalities.

Chlorine and ammonia together constitute the overwhelming proportion of industrially manufactured gases, but chlorine is very much more toxic, and much more widely feared due to its early use as a war gas. As with ammonia there is a fairly extensive record of incidents which have had fatal consequences, extending over a period of sixty years, with comparable totals and a similar spread across the long chain from manufacture through transport to use. Up to and including 1952, there were six reported incidents throughout the world involving catastrophic losses of containment from bulk liquid storage. A total of 115 people died in these incidents. They were associated with mechanical failures which would be very unlikely under modern standards of equipment and operating procedures. There have been no further failures of major storage vessels since 1952, during which time manufacture and use has increased tenfold.

Table 2.1 contrasts some of the published differences between chlorine and ammonia relating to hazards.

The most notable statistic from the reported incidents where there was a major loss of containment is that, with one exception, fatalities are within about 400 metres of the hazard source. The person most at risk appears to be one who is overtaken by a release close by and within a short space of time: someone exposed to very high initial concentrations, perhaps for short periods, but with little chance of escape or relief.

Table 2.1

Concentration contrasts between chlorine and ammonia

	Cl_2 ppm	NH_3 ppm
Odour threshold	around 1	around 5
Bearable for 30–60 minutes	5	250
Lowest reported lethal concentration at 30 minutes for any species	60	5000

There is nothing in the case histories to lend credence to predictions in the original Canvey Report suggesting fatalities from an ammonia release of up to 18 000 people at distances of 10 kilometres and more, albeit at an exceedingly low event frequency.

Montanas' runaway train

The most dramatic catastrophe involving chlorine in recent times involved a runaway train in Mexico.[10] On August 1981, whilst descending a long, three in ten gradient, the brakes failed on a train carrying a large quantity of chlorine between Pajaritos and Tampico. The train's radio was used to warn off other trains on the single-line track, but it eventually derailed on a bend at over 80 km per hour. The pile-up included 28 of the 32 chlorine tank cars, each of 50-tonne capacity. It has been estimated that over 100 tonnes of chlorine gas was immediately released. A gentle breeze took the gas cloud up the valley and through the village of Montanas, with a population of 400. In Montanas station there were 300 people on a passenger train which had sought safety in the spur, following the radio warning.

Seventeen people died in the accident, four of them in the rail crash, and thirteen as a result of chlorine. A thousand were treated in hospital, of whom 256 were detained but recovered. The village was evacuated for several days during the long-drawn-out cleaning up and recovery operation.

It is important to note that in this incident 1000 people were known to be gassed fairly close to the source in a release cloud which must be amongst the biggest ever. The percentage fatality was quite low, which further suggests that some of the well-known societal risk estimates, for example Canvey and Rijnmond, may have been much too conservative. However, this event does make clear the very large ratio of deaths to cases needing medical attention. This may impose a heavy burden upon hospital staff in such circumstances.

Whilst chlorine and ammonia are overwhelmingly important because of the large quantities in manufacture, storage and distribution, other industrial gases may be lethal at far lower concentration levels.

Table 2.2 gives the LC_{50} values for a number of industrial gases. These values are based upon information made available as a result of a research programme at Loughborough University of Technology,[11] but they are subject to review.

The base case is for healthy humans at rest, and is directly comparable to the animal test data found in the literature. The standard case is based on a concept developed by Withers and Lees. It applies to the enhanced breathing following physical activity of humans when, for instance, they run away.

The exponent n refers to the load formula $C \times T^n$ = constant. It enables LC_{50} values to be derived for other exposure times.

Bhopal

Methyl isocyanate is the substance responsible for the greatest industrial catastrophe yet experienced. There have been numerous

Table 2.2
Lethal toxicity levels for some industrial gases

	LC_{50} at 30 minutes, ppm		
Gas	Base case	Standard case	Exponent 'n'
Ammonia	11 500	5 800	$\frac{1}{2}$
Sulphur dioxide	2 300	1 150	$\frac{1}{2}$
Hydrogen fluoride	1 200	600	$\frac{1}{2}$
Bromine	750	375	$\frac{1}{2}$
Hydrogen sulphide	640	320	$\frac{1}{2}$
Chlorine	500	250	$\frac{1}{2}$
Hydrogen cyanide	131	65	$\frac{1}{2}$
Phosgene	53	26	1
Methyl isocyanate	34	17	1

reports upon the events of 2–3 December 1984 when a lethal cloud drifted over an unsuspecting neighbourhood in Bhopal,[12] India, killing at least 2500 people and leading to over 200 000 needing medical treatment. What follows is based upon an agreed account issued after an investigation by an international group of trades union officials.[13]

Large volumes of methyl isocyanate (MIC) were stored in the Union Carbide plant at Bhopal for use as an intermediate in the manufacture of pesticide. Reports on the quantity of MIC held in tank 610 at the time of the accident vary. Its capacity was approximately 70 tons. The Union Carbide technical instruction manual recommends an operating limit of 50 per cent, but it seems that at the time of the accident it was around 75–87 per cent full. According to the Bhopal workers all three MIC storage tanks were frequently filled above the recommended level. But many of the level gauges in the plant were said to be not working or unreliable, and not trusted by the workforce.

According to calculations made by Union Carbide after the accident, tank 611 was about 40 per cent full on the night of 2 December and tank 619 had a small amount of contaminated MIC despite the requirement in the Union Carbide manual that one storage tank always be kept empty in case of emergencies.

Sometime after 9.30 p.m. on 2 December the workers began to flush out several of the pipelines in the plant. This should have been a routine maintenance operation but as a result of some partly leaking isolation valves and, by oversight, the omission to insert a slip plate (which would have formed a barrier in the pipe), and a remotely operated valve which was open when it should have been shut, about 120–240 gallons of water was mistakenly added to tank 610.

The water began to react with the MIC in an exothermic reaction, and the workers close to where the lines were being washed soon began to suffer eye and throat irritation from MIC escaping from tank 610 back along the pipe through which the water had flowed.

By 11 p.m. it was noted in the control room that the pressure in tank 610 had risen from 2 to 10 psi. A runaway reaction had started in tank 610, probably assisted by contamination in the MIC which attacked the walls of the tank, releasing iron which intensified the reaction.

If either tank 611 or 619 had been empty, it could have been used as a surge tank to contain some of the reacting MIC, thus giving the operators more time to regain control of their plant. But since the operators were not sure how much was in these tanks, due to the unreliable gauges, and since they were fearful of spreading the reaction, the lines to the standby tanks were not opened.

At 12.15 a.m. the tank pressure was again noted in the control room; it was 30 psi and seconds later went off the scale. The rupture disc on the tank was designed to give way at 40 psi, and when this happened the contents of the tank rushed through a vent pipe to a scrubber, which was designed to neutralize escaping MIC.

It is not clear whether the scrubber worked that night. The pumps had been shut off and the instruments in the control room indicated that they could not be restarted. In any case the operating manual casts some doubt as to whether the scrubber could have had the capacity to handle such a massive release.

From the scrubber the gas should have gone to a flare tower designed to burn up escaping MIC. But this unit was out of service and the pipe leading to it had been removed some weeks before for maintenance.

Finally, the Bhopal plant was equipped with an emergency water spray system designed to knock down escaping vapours before they could drift across the site perimeter. Several workers made a last-ditch effort to put this into operation. After they had struggled with the poorly maintained valves to turn on the water spray, they found that there was insufficient pressure to reach the gas cloud.

There are no planning restrictions in Bhopal, and after the plant had been built people were allowed to build their living quarters right up to the factory perimeter fence. By 1 a.m. on 3 December a lethal vapour cloud was drifting without warning over thousands of people, mostly fast asleep and totally unprepared to face such an emergency. Very few of them were to escape unharmed.

The authorities too were unprepared, and those offering medical help had little idea what could be done; it was even sometime before the name of the gas was clearly identified, and there was much public uncertainty about the measure of its toxicity.

The main causes of the tragedy at Bhopal seem to be:

1 The quantity of MIC in storage. This was much larger than recommended by the operating instructions. It has been suggested that a more modern process technology would not require even the recommended quantities. Either way this is a failure by the management.
2 The failure by the maintenance and process workers to follow recommended procedures during the flushing-out operation. This is a failure by management at a lower level.
3 The inadequate maintenance of essential pieces of equipment and a disregard of the risks in operating the plant in such circumstances. This is a failure of the management and workforce, at all levels.
4 The poor training and competence of the process workers, including the apparent lack of any training in emergency procedures to provide a considered response to untoward difficulties. Furthermore, practical training in emergency plans would have made clear to all concerned the more vital plant deficiencies.
5 The lack of involvement by the public authorities not only in matters of safety inspection and emergency plans, but so as to inform the public and to refuse to permit housing development in a risky situation.

Perhaps the worst aspect of the Bhopal tragedy is that such lessons had already been learnt over the years in the industrial countries where they are incorporated in the day-to-day management activities of the major petrochemicals companies and the public authorities who look after the interests of the local communities. These activities are often subject to national laws and regulations, and in the European countries the enactment and observance of safety laws relating to possible major hazards received a major stimulus from the 1976 incident at Seveso.

Seveso

TCDD (2,3.7,8,Tetrachlorodibenzoparadioxin) is one of a whole range of dioxins. Many of these are presumed harmless, but TCDD is very dangerous. It appears to interfere with metabolic processes, has an LD_{50} level of 0.6 μg/kg, and has been shown to be fatal to experimental animals in doses as low as 10^{-9} times the bodyweight. There are varying degrees of evidence for carcinogenic, mutagenic, and teratogenic properties, and it can be taken into the body by ingestion, inhalation, or merely by skin contact. A common symptom of mild TCDD poisoning is chloracne, which may take a year to clear up.

Dioxin may be produced by the incineration of appropriate chemicals and chemical waste, and may form an unwanted air pollutant from municipal, hospital, or industrial incinerators. Much depends upon the operating temperature, the turbulence in the incinerator and the residence time, and the production of TCDD in such a manner has yet to be firmly established. The growing awareness of possible dangers has, however, resulted in the revision of some incinerator working practices, notably in the use of higher operating temperatures.

The only confirmed deaths by dioxin poisoning took place in the Netherlands in 1963, where a leak of 0.03–0.2 kg of TCDD occurred at Phillips-Duphar, a pharmaceutical manufacturing company. Some fifty people were involved in cleaning up the place after the leakage, of whom four subsequently died and about a dozen suffered occasional skin troubles. The plant was sealed for ten years and then dismantled from the inside brick by brick, the rubble being embedded in concrete blocks which were sunk in the Atlantic.

Other accidents involving TCDD have been reported by BASF at Ludwigshafen in 1953, and by Coalite Ltd at Bolsover in 1968 and 1974. Although no deaths were positively reported these three further incidents demonstrated the severe problems posed by the necessary cleaning-up operations.

At Seveso,[14] the Icmesa Chemical Company, like Coalite at Bolsover, had a process manufacturing trichlorphenol (TCP), well known as a useful domestic antiseptic. TCDD can be obtained by the elimination of two molecules of HCl from 2,4,5-trichlorphenol, but is usually formed only in very minute quantities in the production process for TCP. Nevertheless, it has to be eliminated in the normal course of events.

The production process involved a batch reactor vessel charged with tetrachlorbenzene, caustic soda, ethylene glycol and xylene. On Friday, 9 July, the batch was started about ten hours later than usual. An initial hydrolysis reaction was completed together with the first of two distillations when the schedule was stopped at around 5 a.m. on the Saturday so that the plant could be closed for the weekend holiday, since the complete operation would take 23 hours. The temperature was then around 158°C, somewhat below the normal operating temperature, so the water cooling was switched off and the stirrer stopped. A later investigation established that an unforeseen runaway exothermic reaction had begun which was to take the temperature to 450°–500°C. At about 12.40 p.m. the 4 inch diameter safety bursting disc on the reactor gave way. Unfortunately, the plant was not designed to discharge into a closed system, but to the atmosphere.

Thus, the contents of the reactor were discharged over the surrounding district in the form of a reddish cloud. The discharge was stopped after about 20 minutes by the foreman who, hearing the discharge making a noise in the vent pipe, applied cooling water at appropriate points to the system.

It has been suggested that 2 kg of TCDD were released, although the normal amount of dioxin present as a contaminant in the crude product and which has to be eliminated by separation followed by high-temperature incineration is only about 5 gm per reactor batch.

Following the release, the company took emergency action to clean up the works. A number of drums of waste material containing TCDD were assembled for disposal purposes and their trans-shipment across national frontiers created considerable media attention and public disquiet throughout the European Community. Works staff attempted to warn members of the local community of the dangers, but had difficulty in making contact with the appropriate authority because of the holiday.

In the immediate area of the release, vegetation was contaminated and animals began to die. Four days after the release the first child fell ill, and after ten days the first evacuation of people took place. Altogether some 750 people were evacuated from an area covering just over one square kilometre. A major exercise to decontaminate the area was started, which was to last a number of years.

Although no one died as a result of this incident, 178 people were treated for chloracne and more than 70 000 animals were slaughtered as a precaution. The total damage has been estimated at £55 million. The legislative consequences of Seveso were significant throughout the EC and are described in the next chapter.

The accident at Seveso has been likened to the situation which could occur after an accident to a nuclear reactor. This might release radioactive iodine-131 to settle over a surrounding area, and people exposed to it would be liable to contract cancer over a number of years.

Nuclear hazards

On 26 April 1986 such an accident was to take place at the Chernobyl power station in the Ukraine, USSR. But since radioactive iodine-131 has a half-life of only eight days it would not be expected to pose anything like the intractable cleaning up problem presented by TCDD. Although radioactive iodine and other relatively short-life substances might be expected to form the greater part of an accidental release from a nuclear reactor, other very long-life materials such as caesium-137 could also be released and they would present a most intractable cleaning up problem.

Three Mile Island

A very serious incident that has been extensively reported was at Three Mile Island PWR (pressurized water reactor) station in the USA in 1979. Largely due to the containment vessel which is intrinsic to the design of a PWR, no one's health was affected, although the owners suffered extremely heavy financial losses. It received much emotive and exaggerated attention from the media, and has proved a source of inspiration to the writers of science fiction. An exhaustive review was provided by the US National Academy of Sciences in 1980, which has provided valuable lessons on the interplay between the human operator and large control systems.[15]

Chernobyl

Although it lags behind the USA and France, the USSR is a substantial producer of nuclear power, having rather more nuclear generating capacity than Japan and about four times as much as the UK. It was one of the first countries to manufacture nuclear power, commercial operation starting in 1954. It exports substantial nuclear generating facilities to the Eastern European countries. These exports are of the PWR type, the system currently most favoured by the international nuclear community. Nuclear energy provides some 11 per cent of the USSR's power. Of this about 40 per cent comes from PWR systems and 60 per cent from 'RBMK' reactors whose design is unique to the USSR.

Because nuclear power arouses such strong feelings any accident receives overwhelming media attention: '70 000 Dead Reds' was a not altogether unexpected headline in the UK popular press when Western countries first became aware of the serious accident at the RBMK reactor at Chernobyl. Even *The Times* headlined 2000 dead on its front page of 30 April, based on a report from the UPI news agency. At the time only two people had actually been killed but, typically, the corrections were to appear somewhat later, if at all, in small print on inside pages.

The RBMK reactor system, like the PWR but unlike the UK Magnox reactor, uses ordinary water to cool the uranium fuel but, unlike the PWR and like the Magnox, uses graphite as a moderator of the nuclear reaction. It may be thought of as a complicated hybrid between the gas-cooled Magnox and the water-cooled PWR.

The Chernobyl disaster occurred primarily due to a lack of understanding by the Russian operating staff, who were carrying out unauthorized experiments with their No. 4 RBMK reactor. A secondary reason was that this type of reactor is inherently unstable under the circumstances induced by the operators.

In brief, the operators decided to test the ability of a turbine generator to power certain of the cooling pumps while freewheeling to a standstill. It was felt that this could be advantageous if the station became disconnected from the grid, and the reactor shut down during a loss of coolant accident.

As the turbine slowed down so did the pumps, and a point was reached where they were no longer pumping enough water to keep the reactor cool. At this point the automatic safety system should have tripped the reactor, but unfortunately the operators had turned this off to save power. Accordingly the reactor began to generate additional steam and this formed extra bubbles and voids in the cooling water. A runaway situation soon developed with the additional heat generating still more steam, with even more voids and even less cooling.

The operators tried to save the situation by releasing the control rods manually so as to trip the reactor. But they were too late, the rapid and massive surge in power had caused the temperature to rise steeply and the increased steam pressure first burst the pressure tubes and then breached the concrete vessel around the reactor. The rupture in the containment of the boiling water system led to a chemical reaction between the water and the zirconium alloy which provides a cladding for the fuel elements. This produced hydrogen which reacted with the air entering the core through the breaches in the concrete vessel to produce a second, massive explosion. This blew off part of the refuelling hall above the reactor and released radioactive material into the atmosphere.

The large initial release of radioactive material from these two explosions was followed by a further release associated with the intense burning in air of the hot and exposed graphite moderator. It was not until 6 May, ten days after the initial accident, that this fire was brought under control.

Such a train of events would not be possible in a PWR because there is no graphite and because of its containment, or in a Magnox because there is no water. However, other untoward reactions are possible in a PWR but these are reckoned to be contained or at least ameliorated in its steel pressure vessel, and the concrete secondary containment, features not provided in the RBMK system. Nevertheless, the full report on the sequence of events issued by the Soviet authorities at a meeting of the International Atomic Energy Agency at Vienna in September 1986 is being read with interest by those involved with the nuclear industry throughout the world.

It admitted that their reactor design had deficiencies but also listed six major errors committed by the operating staff, who were blamed for the tragedy, namely:

1 against all regulations the emergency cooling system had been turned off almost twelve hours before the disaster;
2 power levels in the reactor were allowed to drop below the permitted level causing the reactor to become unstable;
3 the main circulation pumps were overloaded which caused the cooling system to approach saturation temperature;
4 the intertrip between the turbine and the reactor was disconnected;
5 the control rods were withdrawn further than allowed by the regulations and so became ineffective;

6 the safety mechanisms protecting steam pressure and water levels were turned off.

As with the most major industrial accidents the chief casualties were to occur amongst the process workers and firemen, who so bravely fought to contain the fire and leakage of radiation. Later on 26 April 129 foremen and operators were flown to Moscow, suffering from most serious radiation exposure. Another 180 with radiation injury were evacuated the following day.

Medical specialists from all over the world were flown to Moscow to help deploy the most modern methods of treatment known to medical science. Amongst these methods was the transplant of bone marrow although this was not to prove very effective. At the beginning of June the treatment of the 300 most seriously affected had not prevented the deaths of over 20 people with some 30 more in a serious condition. But around 80 of the 300 had been discharged for convalescence. In the September report the official death toll was put at 31 with 203 people suffering from radiation sickness, 22 of them very seriously.

The initial wind direction was such that it was felt that the local population would receive less contamination if they sheltered rather than travelled. Twenty of the first 36 hours were at night and bed is a good place to shelter. The wind then changed and it was decided to evacuate all persons living within 30 km of the accident, a population of some 92 000. This was completed in about three hours. Eleven hundred buses were employed, and 230 medical teams involving 5000 doctors and nurses examined all these people, who will continue to be monitored for a long time to come. A further 250 000 children were subsequently evacuated from the region around Kiev, the Ukrainian capital, for the duration of their summer holidays.

Experts of the International Atomic Energy Agency visited the scene during the first week of May. The increase in the radioactivity in the atmosphere was monitored extensively throughout the Western countries where there was much public concern about safety. Traces of extra radioactivity were detected in Scotland by 6 May. People there were officially advised not to drink rainwater and EC governments endeavoured to impose a collective ban on food imports from the Soviet bloc. Reaction was at its most extreme in Greece where fist fights broke out in a supermarket between panic buyers of the last cans of evaporated milk and bottles of mineral water, because of their fears of radiation from far away Chernobyl.

Meanwhile the International Agency team was generally reassuring about the dangers from radioactivity. It said that most of the released radioactive material was of the short-lived variety, including iodine-131 which made up half of the emission. Potassium iodine tablets were widely distributed for protection of the people against the effects of iodine-131. Radiation levels within the evacuated zone around Chernobyl dropped from an estimated 10–15 mr per hour soon after the explosion to 0.15 mr per hour by 8 May.

It has been estimated conservatively by the Soviet authorities that the 'average' individual among the 75 million people in the USSR will now incur a total extra radiation dose of 3.3 rem over his lifetime as a result of the Chernobyl disaster. Using the risk calculation based on linear extrapolation found in the conservative recommendations of the International Commission on Radiological Protection, it is possible to calculate that such an exposure might result in an excess of 24 000 cancer deaths over the 9.4 million 'natural' cancer deaths which will occur in the USSR during the next seventy years. A less conservative calculation, trying to account more realistically for low-level radiation effects, might predict 2000 deaths instead of 24 000. On the same conservative assumptions, medical use of radiation will involve another 20 000 deaths and exposure to natural radiation will involve 100 000 extra deaths. These low-level radiation effects are subject to considerable uncertainty, however.

At the September conference the Soviet authorities stated that a widespread campaign of measurement of radiation levels had caused them to revise their original estimates and that the actual levels were only about one-tenth of those estimated at the time of the release. The 135 000 people evacuated from the 30 km zone around the reactor will be regularly re-examined medically, but in the light of the dose they have received it is not thought that any special measures will have to be taken by the medical authorities or, indeed, that any anomalous health pattern is likely to emerge.

On 9 May the safety committee of the Nuclear Energy Agency in Paris issued a statement saying that the Chernobyl explosion had caused no significant danger to health in any Western country; levels had nowhere exceeded one-tenth of the internationally agreed safety limit. A government communiqué issued in Warsaw on 7 May said that the average extra accumulated dose of radiation on the body between 28 April and 2 May amounted to 25 mr in Poland which was only five per cent of the 500 mr level considered acceptable to the Polish authorities.

By 2 June Chernobyl had almost disappeared from the front pages of the Western newspapers, while the Soviet papers were reporting good progress in the construction of a concrete tomb around the stricken reactor so that preparations could be made for restarting the other three Chernobyl reactors before the year-end.

On- and off-site hazards including transport

With the exception of Bhopal and San Juan Ixhuatepec, the case histories have illustrated the statistical truth that much greater risks are run by the workers and firemen on factory sites than by the general public in the surrounding district.

The growth of high-density living accommodation so close to hazardous installations, as at Bhopal and San Juan Ixhuatepec, is a development normally avoided in industrialized countries, where it is usual to find an unoccupied buffer zone between a hazardous industrial site and residential dwellings. The width of such a zone does not have to be so great as to cast a 'planning blight' over the district.

The situation is illustrated in the Rijnmond Report,[16] which contrasts the predicted annual death rate between workers on six industrial installations with the surrounding population. The predicted data is reproduced in Table 2.3.

As was mentioned in the previous chapter, the consultants were directed to take a conservative value for the toxicity of chlorine and ammonia. If more recent and realistic figures were to be introduced the following data emerges for annual fatalities for employees and population respectively:

1	Ammonia	4.0×10^{-5};	3×10^{-6};
2	Chlorine	5.0×10^{-4};	1×10^{-4}.

Table 2.3

Predicted Rijnmond annual fatalities

Study object	Employees	Population
Acrynonitrile	2.1×10^{-3}	7.9×10^{-6}
Ammonia	2.1×10^{-3}	2×10^{-4}
Chlorine	1.1×10^{-2}	3.6×10^{-3}
LNG	1.5×10^{-7}	6.8×10^{-10}
Propylene	1.1×10^{-4}	3.7×10^{-5}
Hydrodesulphurizer	1.0×10^{-6}	0

A survey of 162 hydrocarbon vapour cloud explosions carried out by Wiekema provides further evidence from case histories of the greater risks on site.[17] In only three cases did the cloud drift an appreciable distance from the site, and it is probably true that many of the smaller incidents have not been reported upon, decreasing the percentage of total incidents which could affect the population still further.[18]

A modern chemical process plant requires high-calibre staff who are well aware of the relative risks and know that small incidents, although inevitable, could be the precursors of larger and more serious incidents if they do not prompt initiatives to remedy deficiencies. Since their own families will form part of the surrounding community, such people are likely to be well aware of their wider responsibilities and not allow unnecessary risks. However, there is a responsibility on all concerned to see that they are well trained and not merely well paid.

Not all hazards to the general public arise directly from the on-site activities of the petrochemicals industries. Many casualties arise from off-site activities, particularly in transport. Statistics collected by Kletz and Turner[19] over many years show comparable numbers of casualties between transport and fixed installations. Between 1970 and 1979 an average of 147 people were killed annually in accidents at fixed oil and chemical installations, and 101 in transport accidents. The range in fatalities per incident was 4–13 for fixed installations and 2–211 for transport.

Road tanker tragedy

On 11 July 1978 an articulated road tanker exploded as it was travelling along a coastal road adjacent to a campsite near San Carlos de la Rapita, Spain.[20] The tanker had been carrying propylene, and the tanker company stated that the driver was under instructions not to use the road past the campsite, but to use a nearby toll road. It also emerged that the tanker had been overfilled and had no safety valve. A leak developed in the tanker as it approached the campsite and a cloud of white gas started to form. This ignited, probably from a camping stove, and the tanker broke into four pieces. The cloud of released polypropylene ignited as a low-level fireball which initially killed some 150 of the 500 campers but the death toll later increased to 211. Pieces of the tanker travelled up to 300 metres from its initial position and the scantily dressed bathers on the site had no chance of survival in the fierce heat.

The vast majority of hazardous materials are carried in well-maintained road vehicles belonging to responsible companies which operate under strictly enforced protocols in so far as training of drivers, equipment carried, headway, routes, and so on, are concerned. In consequence such vehicles have a much lower accident rate than private motor cars. The UK accident statistics show that even general goods vehicles over 1.5 tonnes have fewer accidents than private cars; in 1981 there were 79 accidents per 100 million miles for 1.5-tonne goods vehicles as against 110 accidents for private motor cars. The US accident rates for goods vehicles are similar, but the statistics reveal[21] that 42 per cent of goods vehicle accidents are due to crashes with private cars and that only ten per cent are due to some mechanical defect. The difficulty with road transport is that, however well trained the driver and however strict the company protocol, an accident may occur due to the uncontrollable activity of a third party. Typical of such unforeseen and often bizarre events is that which occurred on 3 August 1985 on the highway at Checotah, Oklahoma, USA.[22]

Explosions on the highway

Two elderly ladies intended to turn off the highway but became confused at the exit ramp and proceeded instead along the hard shoulder and attempted to get on to the highway again. Behind them was a military lorry carrying a load of 2000 lb bombs and it collided with the ladies' car. The occupants of both vehicles managed to clamber to safety. In the UK in 1981 there were 33 344 accidents involving goods vehicles, but only thirteen resulted in a fire. On this occasion the car caught fire and the fire spread to the lorry. By the time the emergency services arrived all that could be done was to evacuate the 5000 inhabitants of Checotah and wait for the explosions. The bombs were not fused and only seven out of ten exploded, one after the other, rather than all being detonated together. Nevertheless, a huge crater 35 feet across and 27 feet deep was left in the highway and 45 people, mostly firemen and emergency personnel, were injured.

In the UK it has been estimated that of all hazardous freight movements 80 per cent by weight are by road, and 92 per cent of this is petroleum products. Accidents are less frequent with railway trains. Even in the USA, where a lower standard of track maintenance results in 78 per cent of accidents being due to derailment, there are reckoned to be only 34 train accidents every 100 million miles. Of the incidents on American railways involving flammable goods, 70 per cent have been due to loose or defective fittings, connections, and so on. The statistics show a much better safety record on UK railways so far as hazardous freight is concerned. Meslin has calculated that the chance of a rupture to a chlorine tanker on French railways, which transport some 250 000 tonnes of chlorine involving 5500 tanker movements every year, is less than one in a thousand years.[23]

Safer still is transport by boat in open water, where the statistics show only 0.06 accidents per 100 million miles.

Product hazards and waste disposal problems

Public concern is growing about the hazards associated with the end use, storage and waste disposal associated with the products of the chemicals industry. Major disasters have taken place in these areas. One of the worst was in 1971 when 459 people died from eating wheat and barley stolen from the docks in Al Basrah, Iraq.[24] It had been intended as seedcorn and had been treated with mercury compounds. The deliberate adulteration of expensive foods and drink is an age-old problem but can be lethal, as evinced by events in Italy in 1986, industrial methyl alcohol having been added to wine.

It is not surprising that the media-enhanced public reaction is sometimes out of all proportion to the actual risk. Such was the case in Austria when it was discovered in 1985 that a chemical, diethylene glycol, had been illegally added to wine. The chemical was wrongly described by the media as antifreeze, ethylene glycol. Actually, the diethylene glycol was not likely to be as dangerous to human life as the ethyl alcohol present in much greater proportion in the wine, which was itself an antidote to the adulterating chemical. The diethylene glycol was not discovered as a result of any untoward illnesses, or even by an expert wine taster. It was discovered by the tax office because the guilty blender became too greedy and tried to secure exemption from value added tax on the chemical, claiming it was being used as a food.

Explosions in Salford

A number of incidents have arisen due to the storage of hazardous materials in substandard conditions by traders and hauliers. A typical case occurred at Salford, England,[25] in September 1982, where some 2000 tonnes of chemicals, including some 25 tonnes of sodium chlorate, were stored in a warehouse. The warehouse was in a depressed area scheduled for redevelopment, and vandalism had caused fires in nearby buildings which were being demolished. There was also high-density residential accommodation nearby, including tower block flats. Around midnight on 25 September, fire spread into the warehouse followed by explosions, causing the evacuation of 700 residents, broken windows in 520 homes and damage amounting to £1 million. The subsequent investigation led to successful prosecutions for burglary and arson, and the storage company was fined £500 under the UK Health and Safety at Work Act 1974. Close to the explosive sodium chlorate were stacks of titanium oxide, paradichlorobenzene and trichlorobenzene. After the incident the surrounding area was found to be contaminated by traces of titanium oxide for several miles.

The remedies for such situations do not in general lie with careful calculations of risk assessment but rather with strict compliance with legislation relating to permitted storage of hazardous substances. The Salford warehouse would have been a notifiable installation under the UK Notification of Installations Handling Hazardous Substances Regulations 1982 (NIHHS Regulations).

The safe disposal of hazardous waste is a very important matter, but it lies outside the scope of this book. While there have been no reports of death or serious injury to the general public, there have been many incidents which cause public concern, particularly about the use of the countryside for landfill disposals, the pollution of rivers and seas, and other difficulties in the alternative incineration techniques that may be employed.

Legislative matters form the subject of the next chapter.

References

1 Walker, J., *Disasters*, London: Studio Vista, 1973.

2 Netherlands Ministry of Social Affairs and Public Health, *Report of an Inquiry into the Explosion on 20 January 1968 in Pernis*. State Publishing House, 1968.

3 Zuckerman, S., 'Discussion on the problem of blast injuries', *Proc. R. Soc. Med.*, XXXIV, p. 171, 1942.
4 Settles, J. 'Deficiencies of testing and classification', *Annals of the New York Academy of Sciences*, 152–3, p. 199, 1968.
5 *Analysis of LPG incident, Mexico City, November 1984*, TNO Report 8727–13325, Ref. 85–0222, 1985.
6 *The Flixborough Disaster*, Report of the Court of Inquiry, London: HMSO 0113610750, 1975.
7 'Report on Feyzin', *Fire International*, 12, 16, 1966.
8 Irving, D., *The destruction of Dresden*, London: William Kimber, 1963.
9 Lonsdale, H., 'Ammonia tank failure – South Africa', *A. I. Chem. E. Plant Safety*, 17, p. 126, 1973.
10 *The Runaway Train*, I. Chem. E. Loss Prevention Bulletin 052, p. 7, 1983.
11 Withers, R. M. J., *Foundations for simple computer models*, Loughborough University of Technology MHC/86/2, pp. 4–8, 1986.
12 Michaelis, A. R., 'The lesson of Bhopal', *Interdisciplinary Science Reviews*, 10, p. 193, 1985.
13 *The Trade Union Report on Bhopal*, Geneva: International Federation of Free Trades Unions, 1985.
14 Lees, F. P., *Loss Prevention in the Process Industries*, App. 2, p. 883, London: Butterworth, 1980.
15 Michaelis, A. R., 'High technology accidents, unpredictable and inevitable', *Interdisciplinary Science Reviews*, 5, p. 79, 1985.
16 *Rijnmond Report*, Dordrecht, Netherlands: Reidel, 1982.
17 Wiekema, B. J., *Analysis of vapour cloud accidents*, Proceedings of the Fourth Eurodata Conference, Venice, 1983.
18 Badoux, R. A., *Some experiences of a consulting statistician in industrial safety*, Proceedings of the Fourth Nat. Rel. Conference 3B, 1983.
19 Scottish Development Department, *Report of the Public Inquiry into the Shell/Esso Mossmoran/Braefoot Bay proposals*, 1978.
20 'Causes of Spanish campsite disaster', *Fire*, 72, p. 304, 1979.
21 Kloeber, G., Cornell, M., NcNamara, T. and Moccati, A., *Risk Assessment of Air versus other Transportation Modes*, US Department of Transportation PB80-138480, December, 1979.
22 'Bombs away!', *Hazardous Cargo Bulletin*, p. 50, September 1985.

23 Meslin, T. B., 'The case of chlorine transport in France', *Risk Analysis*, 1, p. 137, 1981.
24 Nash, J. R., *Darkest Hours*, New York: Pocket Books.
25 HSE, *The fire and explosions at B & R Hauliers, Salford*, London: HMSO 0118837028, 1983.

3 Public policy and legislation

While it has previously been remarked that the UK is an industrial country with an anti-industry cultural streak, public opinion does in general recognize the benefits industrial production can bring, and public policy seeks merely to contain or mitigate the risks rather than impose impossible constraints upon industry, which might lead to its abolition. Thus, for example, it is never suggested that domestic supplies of gas should be discontinued because an average of thirty-five people die each year from blast or asphyxiation as a result of untoward incidents, or that the manufacture of motor cars be abolished to prevent some 8000 annual deaths from road accidents.

In industrial chemicals production, no one seriously advocates ceasing to manufacture chlorine, even though it clearly presents a possible hazard. The benefits to our way of life from the widespread availability of cheap supplies of chlorine greatly outweigh the disadvantages. One has only to recall the universal afflictions of typhoid and cholera prior to the use of chlorine in water supplies, for example. Nowadays only two per cent of world chlorine consumption is taken up in water treatment, 85 per cent of it reappears in other chemicals. Of these, vinyl chloride monomer takes the leading position, accounting for one-quarter of the world's chlorine consumption and providing a foundation for many of the plastics in everyday use.[1]

Chlorine is mainly manufactured in large integrated process plants using a brine of common salt as the feedstock. The sodium and chlorine atoms in the salt molecule are separated by electrolysis, with electrical energy comprising as much as 70 per cent of the industry's variable costs. To be competitive, therefore, it is equally or even more important to have a cheap supply of electricity as of feedstock. There are significant differences in energy pricing even within countries of the European Community, but the competitive advantage must ultimately lie with those countries whose supplies are based upon nuclear power. In addition to chlorine, alkali (Na_2O), hydrogen and sodium metal are also likely to be produced at such plants. The hydrogen may be taken directly by pipeline to a margarine producing plant, while the alkali may be used as an essential ingredient in the manufacture of rayon and cellophane, aluminium, glass, pulp and paper, and soap and detergents.

Eleven suppliers operating on a large scale, most of them multinational, account for nearly half the world supply of chlorine and alkali. North America and Western Europe provide 58 per cent of world supply, and the worldwide rate of growth is around three per cent a year.

The benefits we derive from the production of these chemicals are plain for all to see. The disadvantages are found, first, in the hazardous waste which the manufacturing industries must discard and, second, in the risk of significant amounts of chlorine or other toxic chemicals being inadvertently released and then inhaled. Inadvertent releases of chlorine may occur anywhere and at any time in the long chain from manufacture to ultimate end use.

Mitigation of these risks requires regulation, which in turn requires standards. To formulate meaningful standards and enforce regulations requires knowledge and understanding which can come only from observation and measurement carried out under carefully controlled conditions. Difficult choices may have to be made, not only in the final balancing of risks against benefits, but also in the acquisition of the appropriate knowledge needed for the risk assessments.

In the quest for information on lethal toxicity, it is unpleasant even to contemplate that such work might require tests involving animals. Yet, in the last resort, tests involving animals must be carried out where there is no other way of obtaining information needed for the ultimate benefit of society as a whole. In many cases this type of needs assessment may be a matter for debate for which no fully agreed outcome is possible. It may be that there is less difficulty in obtaining agreement for applications which relate to industrial gases like chlorine. But even for established materials such as chlorine it is generally the case that environmental standards and safety practices thought reasonable and adequate by our forebears are not so regarded today; and having further regard to new materials there may be no end to these uncomfortable requirements for animal testing.

Nuclear power has developed very fast since the first commercial station was built in 1953, but it still contributes only a small fraction of the world's energy output – a little over one per cent. However, it now contributes substantially to the energy needs of the industrialized countries, which have come to depend on it. For example, it provides about one-third of the electrical power required by the EC countries, where still further expansion is projected. In this, as in other matters, there is a great technological gap between the developed and the developing world.

Most of the world's energy needs are met by the fossil fuels: coal, oil and natural gas. Hydro and geothermal power can only supply about two per cent of world needs, although these are by far the cheapest sources. Some industrialized countries – for instance, Japan – have no indigenous fossil fuel; others – like West Germany or the UK – have it, but it is rather expensive to extract. To obtain electricity from nuclear energy is generally cheaper than from fossil fuel, and the fuel supply seems to be more commercially secure. As a result, some countries have made very considerable investments in nuclear power. In France it currently provides 65 per cent of total electricity supply and is projected to give 75 per cent by 1990. In Belgium it is 60 per cent, in Sweden 45 per cent, and in West Germany 31 per cent. The UK has around 20 per cent, projected to rise to 25 per cent by 1990. Since it is cheaper it provides about 50 per cent of UK electricity needs in the summertime when demand is at its lowest. Nuclear energy formed about 21 per cent of total EC energy production in 1985, and about 12 per cent of its total energy consumption.

The economics of electricity supply have to be reckoned against a very long time scale. A power station takes around ten years from initial authorization to coming on stream. Capital costs favour large-scale constructions, and the charges vary between countries and over the long time span. This makes for difficulties in international comparisons. However, in the UK in 1965 the Central Electricity Generating Board (CEGB) estimated the costs of nuclear power at 0.475d per unit as against 0.53d per unit for coal power. Its 1985 estimates were 2.94p per unit for nuclear power and 4.29p per unit for coal. The CEGB currently imports nuclear electricity from France 25 per cent cheaper than it can itself provide from fossil fuels. Cost comparisons in the other industrialized countries using both coal and nuclear sources show similar findings, except in North America where coal is very cheap.

Nuclear power stations are very clean places and, it may be argued, cause less environmental impact than even hydro power with its huge dams and artificial reservoirs. The record of deaths and injuries in the nuclear industry is far better than either oil or coal. There has been some controversy about the disposal of radioactive waste although the amounts of high-level waste at a nuclear power station are so small in volume that they could be stored on site.

Nevertheless, in spite of these significant advantages, public opinion shows great concern over nuclear power and, since Chernobyl, several countries have changed their plans and others are facing demands to phase out nuclear production of electricity.

The United Kingdom's dual system

In the UK public policy on major hazards is directed towards making industry as safe as practicable means can provide, and separating the public from residual potential hazards by control of land use and of building development. In 1972 the report of the Robens Committee[2] on health and safety at work endorsed a system of dual control of:

1 the location of developments relating to hazardous industry through planning legislation;
2 the operations of hazardous industries through health and safety legislation enforced by a national inspectorate.

Responsibility for the first of these rests primarily upon local planning authorities, that is to say it is a decentralized activity, while responsibility for the second is essentially a matter for a national body, the Health and Safety Commission (HSC) and Executive (HSE). The HSC comprises representatives from employers' associations, trade unions, and local government.

After the Flixborough disaster an Advisory Committee on Major Hazards (ACMH) was set up under the auspices of the HSC. It was representative of local government, the trade unions, industry and academicians. It discussed a wide range of issues, calling up expert testimony as required.[3] The ACMH endorsed the dual approach, and in particular was against any suggestion that a decision on siting should be the responsibility of HSE experts. Its first report stated:

> Planning controls should be applied to the siting of a notifiable installation, to any modification which would convert an existing activity into a notifiable installation, and to any development in the surrounding area. It might be argued that the siting of potentially hazardous installations should be controlled by the HSE, but we hold very firmly to the view that the siting of all developments should be a matter for planning authorities to determine since the safety implications, however important, cannot be divorced from other planning considerations.

In its second report it said:

> Absolute safety was impossible to achieve and some weighing up of advantage and benefit against risk was inevitable. There are occasions where the advantage and benefit of allowing a hazardous development to be introduced into a particular location are sufficient to override the risk.

Thus, it has come to be generally recognized that the local planning authorities might be better placed than the HSE to pass judgement on the balance of risks versus benefits since they would be likely to have more knowledge of local opinion, contribution to local economy, potential value of land, and so on. The authorities are directly accountable to the public who elect them, and they have legal responsibility for formally approving all developments within their geographic area. Their concern is both with matters which may cause an increase in the population in the region of an existing hazard, for example new hospitals and shopping centres, as well as considering applications from industry for new developments which could increase hazards to existing populations.

Local planning authorities do not have to seek advice from the HSE, but advice must be given when sought; however, it need not be followed. The HSE has issued a set of consultation zones to the authorities indicating when it believes its advice should be sought. The zones include:

1 1500 metres for a bulk chlorine storage;

2 500 metres for LPG (liquid petroleum gas) storage;
3 60 metres for a town's gas holder.

If the HSE disagrees with the authority's decision it can recommend that the matter be made the subject of a public inquiry. Past inquiries have sometimes upheld and sometimes negated the planning authority's decision. Even when a public inquiry has come to a clear and definite conclusion, the minister may take a different view. The whole process may at worst generate more heat than light and be very time-consuming, but at best it provides an effective vehicle for public participation and consent in a decision-making process.

The NIHHS Regulations

The Notification of Installations Handling Hazardous Substances Regulations 1982 were implemented in January 1983 following four years' discussion of a draft document issued in 1978.[4] The substances fall into three categories:

1 toxic substances such as chlorine;
2 highly reactive substances such as sodium chlorate;
3 flammable substances such as propylene.

Quantities were given for each of the long list of substances, above which notification to the HSE became mandatory. Following Salford, the limit for sodium chlorate was set at 25 tonnes.

The regulations involved some 1500 installations in the UK and several thousand miles of pipeline. In addition to chemical process sites, the regulations cover nuclear sites, military installations, explosives factories and mines. The regulations do not, however, cover hazardous substances in transit even when temporarily parked in a warehouse. Each site operator is required to notify the HSE area office (there are special offices for mines and nuclear plants). The information is then transmitted to the Major Hazards Assessment Unit (MHAU) within the HSE, which makes an assessment of the possible impact of the notified installation on its surroundings and fixes an appropriate consultation distance. The consultation distance and the original notification are then passed, via the area office, to the local authority and local fire brigade.

While the NIHHS Regulations were an important step in changing UK legislation affecting major industrial hazards, by the time they were implemented further wide-reaching proposals for legislation were being discussed as a result of Seveso and initiatives being taken by the European Commission in Brussels. These discussions highlighted the requirement for some agreed form of quantified risk assessment.

Agreement on quantification has proved difficult to obtain. This is because precise quantification has inherent difficulties, explained in later chapters, and industry has often expressed reservations through manufacturers' associations such as CONCAWE (Oil Companies' European Organization for Environmental and Health Protection) or CEFIC (Chemical Companies' European Federation).

The CIMAH Regulations[5]

Until recently a company proposing to construct a hazardous installation in Britain has not been required to formulate a quantified risk assessment or even to present a safety case, but under the CIMAH Regulations, made in 1984, such a case must now be presented to the HSE three months before commencement of an activity involving significant inventories of specified hazardous materials. Schedules to the regulations list a number of substances and the quantities involved, whether in industrial processes or in isolated storage, which will involve a statutory obligation under the regulations. Included, for example, is methyl isocyanate, the substance at the heart of the Bhopal disaster. One tonne of this substance is enough to activate the various statutory obligations.

The statutory obligations fall into five broad categories:

1. to report relevant industrial activities to the HSE;
2. to prepare on-site emergency plans;
3. to assist the local planning authority to prepare off-site emergency plans;
4. to demonstrate to the HSE safe operation of the relevant activity;
5. to notify the HSE of major accidents.

The regulations do not apply to any nuclear installation, defence establishment, munitions factory licensed under the Explosives Act 1875, mine or quarry, or any site operated by a waste disposal authority licensed under the Control of Pollution Act 1974.

These regulations implement an EC Directive on Major Accident Hazards and are known as the Control of Industrial Major Accident

Hazard (CIMAH) Regulations. Their implementation will focus attention on 208 sites throughout the UK (but not in Northern Ireland).[6] A much wider range of hazardous installations is subject to notification under the NIHHS Regulations, although some installations fall within the scope of CIMAH but are not notifiable under NIHHS. There is also public pressure in the UK to increase the number of notifiable sites. There are, for example, new requirements to put up a special warning notice or sign outside every installation containing more than 25 tonnes of hazardous material – this involves 34 000 sites. Some people regard the 208 CIMAH sites as the tip of an iceberg.

There is much to be said for insisting that the same type of safety case required under the CIMAH Regulations should also be provided when applications for development of notifiable installations are made. The information which has to be provided for the safety case is specified in the CIMAH Regulations and is explained further in the guidance notes to the regulations. It does not have to be presented to the planning authority or to the public and is a confidential document. The information falls mainly into two categories: the first describes the management structure and procedures for ensuring that the process plant is well controlled in so far as hazards are concerned; the second describes the hazards and the related risk assessment. The precise nature of the required safety case is not specified and in particular it is not stated whether the HSE will accept qualitative arguments or demand quantitative estimates. The notes state: 'Whilst it may be possible for manufacturers to write a safety case in qualitative terms, HSE may well find it easier to accept conclusions which are supported by quantified arguments.'

Among the various European countries there is a degree of non-uniformity in the response to the Seveso directives. Some countries have begun the implementation faster than others, and in some countries the risk assessments that are presented to the regulatory authority are entirely qualitative and secret, whilst elsewhere quantitative assessments and openness are encouraged.

For large projects, planning authorities have sometimes requested some form of environmental impact assessment and, within this, some form of safety case. Problems have tended to arise with such studies for three main reasons. There has been considerable variety in the scope and structure of the studies, the degree of completeness has varied widely, and the authority may have had little experience in evaluating such matters.

The execution of a project is a phased process and at the stage when outline planning permission for a new installation is requested the design is usually only an outline appropriate to the feasibility phase of the project master plan. Nevertheless, such an outline plan should be sufficient for the purposes of a risk assessment. The information required concerns the main process and storage units together with terminal facilities, inventories and operating conditions. This would normally be sufficient to allow the identification of potential releases of hazardous materials for the risk assessment. This type of generic information, largely common to all similar process plants, was used in the Canvey studies even though the plants were already in operation, presumably because it was much more readily available within budgeted time and cost than plant-specific data would have been.

While such a risk assessment based wholly on generic data will be appropriate when seeking outline planning permission, it seems probable that if there is a public inquiry a fuller assessment will be demanded. There will normally be an appreciable period between a planning application and an inquiry and this could give the company time to update its safety case as appropriate. It is most desirable that the structure of such a safety case should conform to the structure of all such cases, whether open or confidential, previously submitted or not.

Any case prepared for a public inquiry must necessarily be an open safety case. While it will be limited to the effects from the company's installation, it may be that the boundaries are wider than those of the outline case. The normal procedure in a public inquiry is to hear evidence by party rather than by issue, with the appellant having the final right of reply. This results in the safety evidence being interspersed with evidence on other topics, and not taken in its own logical order. Since there is no established pattern for a risk assessment, it is often difficult to follow the thread of the arguments presented in the published proceedings, particularly where the technical evidence in a risk assessment is concerned. In big inquiries, however, hearing may be by issue. This has been the case at the Sizewell inquiry.

The dialogue between industry and the local planning authority during the planning application is important and the presentation of open safety cases in a uniform style would help authorities to become familiar with the subject. A proposed standard structure for risk assessments, whether the safety case is open or not, is outlined in Table 3.1.[7]

Table 3.1
A structure for risk assessment

System definition
Definition of system and system boundaries (e.g. inclusive of transport)
Description of installation (inclusive of processes, inventories, vessels)
Release patterns
Statement of causes of release, including causes considered but not analysed (e.g. aircraft crashes)
Estimate of frequency and magnitude of releases (data sources and methods to be referenced)
Emergency scenarios
These will comprise chains of events from emission through dispersion to damage to property and people
Estimates of frequencies and chances down the chain (data sources and models to be referenced)
Statement of damage and injury relationships (models and population data to be referenced)
Presentation of results
Individual risks on and off site, e.g. using plot plans, risk transects, and contours
Societal risks on and off site, e.g. giving accumulated annual fatality rates at various locations, and frequency number (*f*/*n*) curves
Statement of criteria against which risks may be judged
Relevant comment (e.g. sensitivity analysis)

The main purpose of the proposal for a structured safety case is to improve the clarity of the presentation of the risk assessment issues, and to avoid the surprise emergence of an alleged hazard which may have been neglected. Table 3.1 outlines a structure for a complete assessment, inclusive of individual and societal risks. In some cases the assessment may be terminated at an intermediate point in the chain, such as after the estimation of frequency and magnitude of the releases.

The risk assessment is based on the assumption that there are adequate management systems to enforce control of potential hazards. The role of management, which is discussed in the next chapter, operates at all levels in an organization and is not merely concerned with maintaining effectiveness and efficiency within the confines of the plant. Nowadays attempts must be made not only to reconcile the needs of individuals to corporate goals but also to reconcile the needs of an organization to those of the community it serves. The enforcement role of the HSE's inspectorate is, however, primarily focused on matters which bear directly upon the health and safety of the people employed on site.

The complexity of process configurations in a modern chemical process plant makes it difficult if not impossible for any outsider to carry out detailed inspection of the minutiae of equipment systems and control procedures so often responsible for the initiation as well as the prevention of hazards. Of course detailed inspection after an event is another matter, but in the meantime the inspector's role is likely to be most fruitful if it is regarded primarily as consultative and an audit of principles.

It is rare in the UK process industry for disagreements to be so intractable or for issues to be so complex as to warrant a Public Inquiry; nevertheless they do happen and it is as well to understand the system involved, which is unique to the UK.

Public Inquiries

Twenty-eight days before the opening of a Public Inquiry the local planning authority is required to produce a written statement. Only the authority has this obligation. The lack of any requirement for the applicant to provide a written statement prior to the inquiry has been much criticized. It is most desirable that the principal parties to an Inquiry provide written statements in advance of the Inquiry so as to assist all concerned to formulate their evidence.

The local planning authority will always be a party to a Planning Inquiry. There is no general requirement that other government bodies such as the HSE should appear. However, it is highly desirable that both the company and the HSE should appear when requested so that all concerned may question them on their written statements. At the present time where a government department has expressed in writing the view that the application should not be granted or granted only with conditions and the authority has included this view in its written statement, the applicant is entitled to have a representative of that department called by the authority as a witness.

The Secretary of State may appoint an assessor with safety expertise to assist the inspector. He advises the inspector and may examine witnesses, but is not examined himself. He may provide a written report for the inspector who may quote it verbatim in his report. The function of the technical assessor appears satisfactory and should not be changed.

The part played by the HSE in a public inquiry needs to be seen in the context of its wider and ongoing role in the control of hazards. If this role is to be safeguarded, it is desirable that as far as possible it should act as an adviser rather than as a protagonist. At the outset of each Inquiry the HSE should be invited to state its role, the legislation under which it operates, its powers under the legislation, and the limits of such powers. It should define the set of principles by which it judges the nature of the risk and the relevant safety criteria required by the Health and Safety at Work Act, as enforced by the HSE.

It is uncertain how far the HSE will reveal its thinking on the use of risk criteria at an Inquiry, although at the Sizewell 'B' Inquiry the Nuclear Installations Inspectorate (NII) (it is part of the HSE), did make its position clear at the request of the inspector.[8]

NII's written evidence to this Inquiry included the following passages:

> 'No person outside the power station should receive more than one-thirtieth of the 'maximum' radiation dose for the general public (at present 0.5 rem a year) from normal power station operation. This would involve a one in a million chance of developing a fatal cancer for each year a person was so exposed.'

> 'As far as a major accident is concerned such as to lead to an eventual 100 cancer deaths, the NII believes it would be reasonably practicable to reduce the chances to less than one in 300 000 a year. Even if there were 100 reactors operating in the country this would mean that the chances of one event of this kind would be less than one in 3000 a year. For comparison, the chances of a railway accident in Britain killing 100 people are probably of the order of one in 20 per year.'

> 'A more severe accident with the potential for causing 1000 deaths, some immediate, most later, can in the view of the NII be reduced to no more than one in a million per year. The design of the station and its containment should aim at chances of no more than one in ten million a year for such catastrophic events.'

> 'A large-scale release of radioactivity would also be likely to cause widespread contamination of land and buildings for some miles downward from the station. The chances of such an environmental disaster are estimated at less than one in 300 000

a year. By comparison the chances of a tidal surge too large to be controlled by the Thames flood barrier are estimated at one chance in 1000 a year: and in addition there is an unquantified risk of the barrier failing to operate when required.'

'The HSE would not consider that the risks set out in this paper would be such as to lead them to refuse a licence for the Sizewell 'B' power station, provided that they were satisfied that all that it was reasonably practicable to do to reduce risks had in fact been done.'

The Sizewell 'B' Inquiry was in every sense a 'big' Inquiry with very large national issues at stake. Much smaller, perhaps more typical, but certainly concerned with principles of great interest, was the Pheasant Wood Inquiry.[9]

Particularly in the north of England, where there is much need for urban renewal often in locations where there is a long manufacturing tradition, the conflict between safety considerations and economic and social development is most acute. Heavy concentrations of industrial production are often to be found in close proximity to high-density population areas. The Pheasant Wood Inquiry into proposed housing development near to a chlorine and phosgene storage plant owned by ICI arose when the Secretary of State was asked by the HSE to call in the proposal for review after the local planning authority had indicated that it would grant planning permission for the development of a private housing estate, against the firm recommendation of the HSE.

The HSE was probably acting to maintain a principle and to uphold its policy to 'stabilize or reduce' the population density around a perceived hazard. It was opposed to any further development within a zone extending to 1 km from the chlorine plant.

No incidents at the plant had ever led to any member of the public requiring treatment or suffering lasting effects. The plant was an important source of employment in the district and ICI had a good reputation, being seen as the chief industry of the locality. Within a 1 km radius of the main chlorine storage there were already (by 1981) some 3750 homes, an old people's home and three schools. The new application was for the private development of a further 452 homes, shops, a school site and a link road.

At the Inquiry, the developer retained the services of independent consultants who produced an alternative risk assessment to that of the HSE. ICI, although not directly involved, came forward with its own assessment, largely based on past operating performance records.

HSE's assessment came under considerable criticism not only from witnesses but also from the assessor. The appropriateness of its dispersion model, the conservative nature of the input data and the lack of a societal risk estimate were particularly stressed.

Following the Inquiry, which lasted two weeks, the inspector recommended that planning permission be granted. The Secretary of State's decision came eight months later, supported the inspector's recommendations and represented a serious questioning of the HSE's 'stabilize or reduce' policy. However, as a result of the Inquiry and the inspector's recommendations, some re-siting of the school, shops and housing did take place to increase somewhat the separation distances from the ICI plant.

Policy positions of public authorities

There is considerable variation in the rate at which the Seveso directives are being implemented in the member countries of the EC. This is partly a reflection of the differing bureaucracies and parliamentary processes and partly because of differing policies within the member countries in the light of uncertain information.

The countries which have found it possible to make the most open progress with procedures involving quantitative risk assessment are the Netherlands and the UK. In Chapter 1 the use of the *f/n* criteria established by the Provincial Waterstaat Groningen was outlined. It may be that current doubts about the value of quantitative risk assessment are illustrated by these Groningen lines (Figure 1.5), since the gap between the implied total deaths per year of the upper unacceptable line and of the lower acceptable line is a factor of 10 000. The rather wide margin in between is where further assessment is required and where acceptability becomes a matter for debate.

In the face of such uncertainty it may be thought necessary to adopt a somewhat pessimistic position on public policy, which determines what should be accepted. Thus, although in the Netherlands the risk estimates are based on realistic figures for instant deaths to healthy people, the Groningen 'unacceptable' line is rather onerous. While the revised societal risk data given in the second Canvey report appears above the 'unacceptable' Groningen line in Figure 1.5, the inspector at the Public Inquiry found the Canvey risks to be acceptable.

In the UK the regulatory authority tends towards a conservative approach when formulating the risk estimate. Thus, the Canvey data shown in Figure 1.5 was mainly dependent upon LC_{50} data for ammonia, which would be further revised today because of the higher levels of concentration now known to be appropriate for lethal inhalation of ammonia. However, instead of LC_{50} values of instant death to healthy people, an 'HSE dose' is used by the MHAU to define safety distances. This inclines more to the lower confidence level of an LC_{01}, rather than to an LC_{50}, to take into account the uncertainty factors and the more vulnerable elements of the population. Unfortunately even the uncertainty levels at such low levels of damage are difficult to agree and in the opinion of some sections of industry, notably the chemical process industry in Germany, may lead to unnecessary alarm and disquiet among the general public.

The general public has, however, come to accept the principle of scientific determinism, that every event must have a cause, even if it may seek to blame someone else when things go wrong. Risk assessment is not the only form of reassurance that can be given to the public. Another is a clear perception of a strong safety policy, executed under an effective and experienced management. This aspect leads into the subject of the next chapter.

References

1 Brennan, C. K., *The World Chlor-Alkali Outlook 1975–1990*, 1985 International Chlorine Symposium, London: Soc. Chem. Ind./Ellis Horwood, p. 20, 1986.

2 *Report of the Robens Committee on Health and Safety at Work*, Cmnd 5034, London: HMSO, 1972.

3 Health and Safety Commission, *Advisory Committee on Major Hazards*, First, Second, and Third Reports, London: HMSO, 1976, 1979 and 1983.

4 Notification of Installations Handling Hazardous Substances Regulations, SI 1982/1357, London: HMSO.

5 European Community Council Directive 24 June 1982, 'Major accident hazards of certain industrial activities', *Official Journal of the European Communities*, L 230/1, February 1982.

6 Health and Safety Executive, *Guide to the Control of Major*

Accident Hazards Regulations, London: HMSO, 1984.

7 Petts, J., Withers, J. and Lees, F., 'Expert evidence at inquiries into major hazards', *Project Appraisal*, 1, p. 3, 1986.

8 Sizewell 'B' Power Station Inquiry, Documents NII/S/83 (SaF) and NII/S/92 (SaF), July 1984.

9 *Public Inquiry into an application to Wyre Borough Council by Brosely Estates Ltd for development of the Pheasant Wood site*, Report of Inspector S. Reese, Department of Environment PNW 5063/219/16, Manchester, 1981.

4 Management at all levels

In previous chapters the necessity for an effective system of safety management has been stressed. Such a system needs to be seen in the context of the company-wide management organization of which it must form an integral part.

It is often the case that the manufacture of industrial chemicals likely to provide a major hazard is carried out by large organizations, frequently trading on an international scale. In large industrial companies the organization is usually divided into a number of groups each corresponding to a main activity of the business. Each principal group is made as self-sufficient as possible so that the individual in charge has the necessary resources and authority to achieve the business results required. These operating groups form the mainline activities of the company, and the predominant organizational relationship between individuals within this mainline hierarchy is known as a line relationship. A line relationship is generally recognized as the relationship between a senior and his direct subordinate involving the delegation of responsibility and executive instructions downwards and the channel of accountability upwards.

In large organizations, in addition to the mainline groupings, there are frequently found a number of specialist staff departments. These departments often employ specialists who have responsibilities for securing propriety of practice in various key functions throughout the whole organization. Such specialists may operate from their own organizational units at all levels in the company from headquarters downwards. The specialist departments may employ large numbers and they in turn are governed by line relationships within their departmental boundaries. Both the growth of technology and of multinational working appear to foster increases in the number of functional and specialist groups in the larger companies.

In contradistinction to line relationships, functional or specialist relationships are of an indirect nature. There will be an accountability upwards towards higher management for promoting effective performance in the specialist function, but there is often no direct authority over people in the line. The relationship between staff function and line is most often an advisory and service relationship where the specialist is called upon to establish practices and procedures, but which may well be incorporated into the line relationships as mandatory instructions. Occasionally, the line management may feel that there is a conflict between the functional advice or procedure and the line's business requirement. Systems for resolving such conflicts and making clear to all concerned the part each has to play are necessary but it is usually the case that, in the first instance at least, the line management has the right to act as it thinks fit.

It is vital that these responsibility relationships be properly understood in matters affecting safety and where action may have to be taken in an emergency to remedy unforeseen difficulties.

A characteristic of organizations with substantial line and functional activities is that they are usually based upon the concept of a continuing steady state, and that staff in both activity groups tend to accumulate with time. To avoid excessive overhead costs top management will not allow any activity to be undertaken that can be obtained more readily and more cheaply as a service from outside. Thus, it may be the case that maintenance work, not only in respect of hardware items but also in respect of software systems, may be contracted out to specialist suppliers. These specialist contractors may be quite small companies with limited resources who are themselves dependent upon specialist services obtained under contract. They may be required to work to special safety codes established by their customers' specialist staff.

All technologically based companies have to face the continuing problems of technological obsolescence and the need to keep abreast of recent developments to remain competitive. Efforts are usually made to contain the demands of technological renewal to a series of small projects that can be handled within the constraints of the steady-state organizational establishment. However, occasions will arise when a large project will emerge raising special problems for the line and function groups within the company. Since much of the work is non-repetitive, varying efforts and commitments are required from people who may come and go.

In these circumstances, it is commonly found necessary to appoint general managers for particular projects to act as a focal point for all project activities. In large organizations, special project groups may emerge who are line groups in their own right, with special steady-state responsibilities for the management of projects even though each project is transient. Inevitably such project work will involve the further use of outside contractors.

The manager of a large project may therefore be in charge of a multidisciplinary team drawn from a number of line and function groups. He will also depend upon the contributions from contractors who will have separate functional relationships. Some of these people may have responsibilities in more than one project and be accountable to more than one manager or to more than one client.

This trend towards the use of contractors coupled with the web of responsibility and accountability relationships which characterizes the large technologically based company has led to a widespread acceptance of written operating procedures, job descriptions, technical specifications and codes of practice, which are mandatory for both internal and external suppliers of services to the company. While individual organizations may have their own specialist requirements for capital equipment, or have historic associations with particular suppliers who may wish to manufacture to their own house style, the present-day trend is increasingly towards national and international codes of practice and specifications. Designs are often interpreted as computer-aided packages based on standard software.

Table 4.1 illustrates some of the differing attitudes and perspectives that have to be reconciled in the management organizations of large technologically based companies.

Participative control

Modern technology-based process plants require very few people in direct control over the operation at any one time. But the capital investment charges and the monetary value of the throughput per

Table 4.1
Perspectives in management

Phenomena	Transient view	Steady-state view
Work	Finite and non-repetitive	Tends to be self-perpetuating
Work-flow	Depends on task	Authority and work tend to go together up and down the management line
Staff	Come and go. Team composition varies with time	Numbers tend to accumulate if all goes well
Objectives	Multilateral, with control over time and cost uppermost	Unilateral to serve main aims of the business
Relationships	Peer to peer. Authority is based on knowledge rather than position or status	Superior-to-subordinate relationships are the most important
Organization structure	A web of activity and responsibility relationships exists	Line managers have the responsibility for achieving results. Staff advises

process worker are very much higher than for other sectors of industry. The high capital intensity may necessitate continuous operations with the workforce deployed in shifts to provide 24-hour cover, perhaps for seven days a week. A typical arrangement might be four sets of 8-hour-shift teams each of some forty people, working to a schedule which provides rotation of night work and allows a reasonable amount of time off during the day, at regular intervals. There would also be office, maintenance and other support services working on a daily basis. Altogether there might be around 200 people on site during the daytime.

The forty-strong shift team would comprise specialist skills to supervise all aspects of the processes and to provide first-line maintenance of instruments, mechanical and electrical equipment, and so on. While the shift manager would have direct line responsibility for all activities in his team and have line accountability to the factory superintendent, some of the specialists in his team would have functional relationships to others, the chief electrical engineer or the works chemist for example.

The heavy responsibilities carried by such relatively small numbers working regularly together tends to encourage a sense of interdependence within the shift team and a participative style of management. The shift manager will usually seek opportunities to share his thinking and decision-making with his subordinates and to make frequent contact with them. Very often the processes are interlinked with a high level of automatic control and the operation is sufficiently complex that more than one subordinate has to contribute the required technical information and response to a given situation. As a consequence there is a greater need for co-operation and participation in the shift management than was the case when technologies were relatively simple and the chief could be expected to 'know it all' and act accordingly.

During the last twenty years or so, participative management has not lacked enthusiastic advocates anxious to proclaim its virtues and develop its use. The social scientists, almost without exception, propound its advantages in terms of increased productivity based on greater job satisfaction. However, a participative style of management in the process industry is not part of an industrial relations climate where everyone is encouraged to be creative to the extent of 'doing their own thing', neither does it involve group discussions to make things up as they go along in an emergency. Preconditions for participative management in the process industry include a high degree of order and discipline within the company, strict observance of operating procedures and manufacturing specifications, extensive operator training, planned preventive maintenance and the employment at all levels of appropriately qualified staff who are subject to regular performance reviews.

Compliance with codes, procedures and regulations

Most industrialized countries have national rules and laws governing safety at work expressed in terms which enable easy verification by a visiting inspector. Thus, they may require some form of guard around rotating machinery, or handrails to prevent people falling. Quantification is usually confined to aspects easily verified, for example in the UK handrails to stop people falling 6 feet or more must be between 3 feet and 3 feet 9 inches high. More complicated matters are often less precise. Under UK legislation both employees and employers have duties to control hazards and to minimize risks. While the HSE has enforcement powers to deal with infringement of the legislation, the national regulations do not aim to secure absolute safety, rather to seek a reduction in risk 'as far as is reasonably practicable'. Thus, safety is often secured by 'the best practicable means'. Other countries have legislation which attempts to go further than the UK and establish absolute safety standards by quantitative means. It is true that dual standards operate in the UK since a plant built on a new 'greenfield' site will often be able to achieve higher standards and lower risks than an existing plant on an old site. Yet there is little evidence in the reported industrial accident statistics amongst the industrially developed countries to suggest any marked differences in safety practices in fixed chemical process installations.

There can be little doubt that a positive influence has been exerted both by the activities of the professional engineering institutions and by the larger petrochemicals process companies. Both have well-established international links with fraternal organizations. The former have been active in establishing codes and recommended practices whilst the latter have usually established their own sets of rules and regulations which meet and extend the scope of national requirements. Both encourage their members and staff, respectively, to share their working experiences at international conferences.

A frequently quoted example of such activity is provided by the API group in the USA. The group's work includes:[1]

1 design and construction;
2 guide for venting atmospheric and low pressure storage tanks;
3 practice against ignition;
4 guide for inspection of refinery equipment.

Over sixty API standards and recommended practices have been listed. This work has involved the co-operation of engineers from many disciplines and will be of help to any engineer with loss prevention responsibilities in a plant using inflammable liquids.

The DOW Chemical Company of the USA publishes a set of about ninety loss prevention principles,[2] categorized under twelve subject headings. These form a set of recommended practices, which are constantly reviewed and updated, and which are to be observed at all phases and management levels in its companies' activity. Their scope ranges from general topics of plant layout to such detailed topics as instrumentation and computer control. Similar corporate activity is to be found in most of the large chemicals companies which also co-operate with each other either directly or under the auspices of recognized manufacturers' associations.

Thus, the Chlorine Institute in the USA and the Chemical Industries Association (CIA) in the UK have developed details on design of tanks to minimize the frequency and extent of leakage in bulk handling in customers' installations as well as emergency procedures and equipment to handle leaks of toxic material.

In the UK, ICI and other companies, in addition to their own internal systems and procedures and their support to the CIA, participate in schemes such as 'Chemsafe' and 'Chlor-aid'.[3] The basis of the 'Chlor-aid' agreement is the formation of a common emergency organization, equipment, and techniques between four manufacturers/suppliers and a handling/packer company as follows:

1 Associated Octel;
2 British Oxygen Company;
3 Hays Chemicals;
4 ICI;
5 Staveley Chemicals.

The scheme enables the nearest and most appropriate company to respond rapidly to a chlorine incident involving any of the containers used by any of the companies. Although the 'Chemsafe' scheme covers chemicals in transit only, the 'Clor-aid' scheme covers chlorine incidents at customers' premises.

Another important influence stems from the activities of the insurance companies. These seek information when setting premiums and develop their database as a direct result of paying policyholders' claims. Claims from fires and explosions are of particular interest. Insurers naturally wish to minimize their losses and issue all kinds of loss prevention data sheets, bulletins and recommended practices. Their staff frequently attend and play an important part in the international conferences staged by the professional institutions.[4] They can also play a more direct role by making the manufacture of specified pressure vessels in accordance with a recognized code a condition of acceptance of a particular contract. Thus, for example, an insurer may well demand that new equipment be designed and supplied to comply with and bear the authorized stamp of the ASME Code, Section VIII, Division 1.

A manufacturing requirement: the ASME Code

From the onset of the Industrial Revolution reputable manufacturers of pressure vessels have worked to well-defined engineering standards of their own. These have been incorporated in company manuals often jealously guarded and only issued on a confidential basis within the company. These company manuals might also lay down control procedures for design, manufacture and inspection, the issue of drawings, change orders, and so on. Such companies would also operate protracted and rigorous apprentice training schemes, and qualified engineers would be expected to have practical experience in all phases from initial design through drawing to manufacture, erection and commissioning. The work was often arduous: back breaking in the drawing office, energetic in the works, whilst erection and commissioning often took place in far away places under unfavourable conditions and during unsocial hours. Under such circumstances company procedures might be more honoured in the breach than in the observance, with an engineer being expected to remedy deficiencies or counter unexpected difficulties as best he could. Nevertheless, there were compensations and such companies often had a favoured status with their customers whose own engineers were usually apprentice-trained with the manufacturer and who naturally preferred to see further orders placed with the source they knew best.[5]

In the period after World War 1 common engineering standards for many constructions were agreed on a national basis by various institutions and the manufacturers were encouraged to construct equipment in accordance with these standards, but usually still working to their own procedural codes and management systems.

Such arrangements served plant manufacturers in the UK pretty well until the period after World War 2. In the 1950s the UK was still supplying some 25 per cent of the world's exported plant. By 1986 this had dwindled to less than 8 per cent, and appreciable amounts were being imported. The reasons for this change are complex. Many user companies had established their own engineering capabilities enabling them to specify and commission their own plant. Specialist design and construction companies had also become established, having an international capability and a full range of engineering functions. These have widened the potential sources of manufactured equipment, not least from the newly industrialized countries who may be much cheaper. Over the same period the plant operation became increasingly complex, increasingly capital-intensive and with a rapid growth in the size of the unit operations. Increased technological complexity and higher capital charges placed a greater premium on reliability and the integrity of the initial design and fabrication; things had to be correct *ab initio*. All this led to an increased demand for national and international standards and codes with which the worldwide manufacturers would be expected to comply and which could be incorporated into the purchasing specifications required for competitive tendering.

The most universally accepted construction code for pressure vessels is administered under the auspices of the American Society of Mechanical Engineers (ASME). Vessels which comply with this code may only be designed, manufactured, modified, or repaired by authorized manufacturers and contractors. Originally limited to the USA and Canada, manufacturers are now authorized throughout the world. While it should not be too difficult for a reputable and long-established UK manufacturer to obtain such authorization, they have been slow to apply and few in number compared with Japan.

One such UK company, Fletcher & Stewart of Derby, first established in the early part of the nineteenth century, obtained its certificate of authorization in 1981; it is shown in Figure 4.1.

ASME could not possibly itself inspect all the worldwide manufacturers and repairers working to its code. Instead it has evolved a set of standards and a way of management audit which governs the integrity of the activities of the manufacturers it has authorized. The principles employed are of general interest and application to risk management, while the subject of pressure vessels is directly relevant to hazards since their possible failure so often provides the top event in a major hazard scenario. The system is conveniently explained by selected illustrations from the Fletcher & Stewart case history.

Management under the ASME Code

The ASME Code is administered by the Boiler and Pressure Vessel Committee set up under the auspices of the American Society of Mechanical Engineers. This meets regularly to consider requests for clarification and extension of its scope and rules. These may be summarized as follows:

1 Rules to cover minimum technological requirements for pressure vessels under the headings of:
   (a) design (including approved materials);
   (b) fabrication (including approved welding competence);
   (c) inspection (including custody of required documentation);

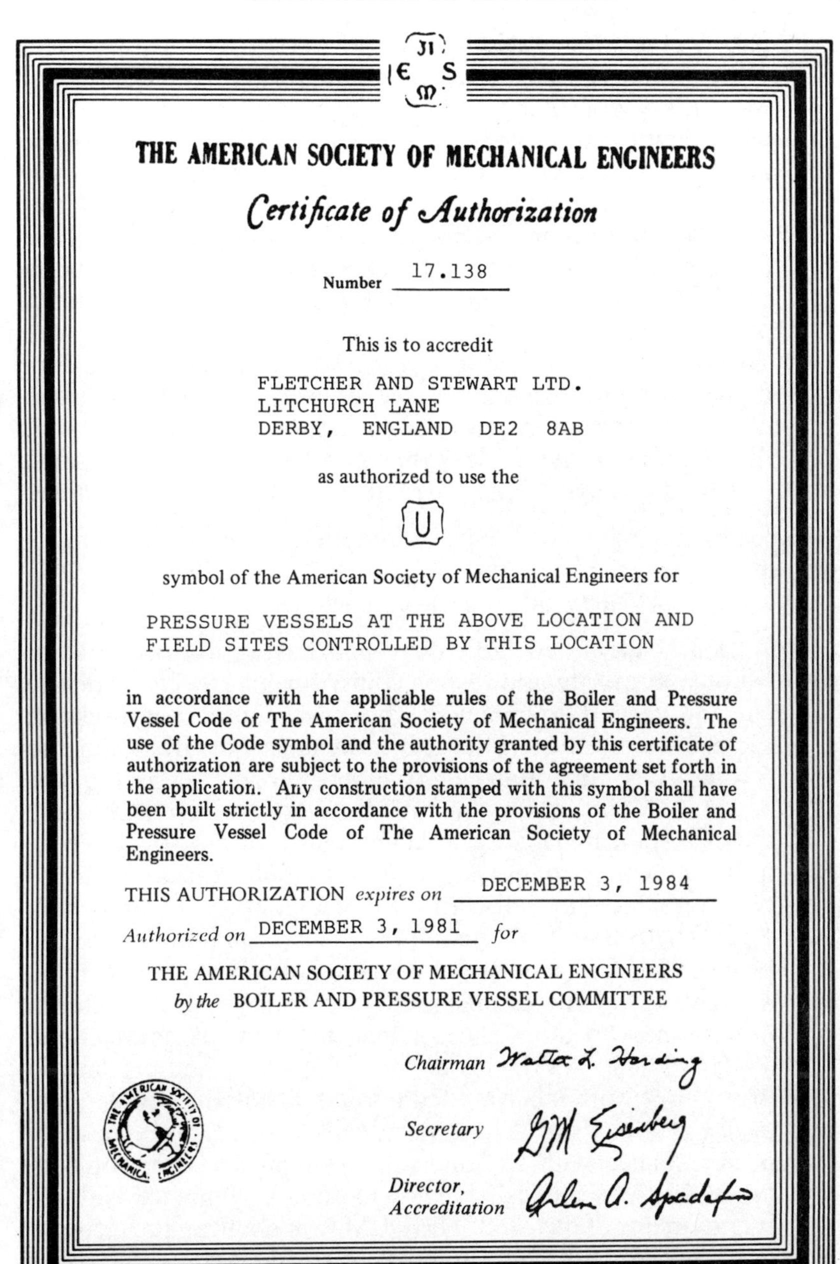

THE AMERICAN SOCIETY OF MECHANICAL ENGINEERS

Certificate of Authorization

Number 17.138

This is to accredit

FLETCHER AND STEWART LTD.
LITCHURCH LANE
DERBY, ENGLAND DE2 8AB

as authorized to use the

U

symbol of the American Society of Mechanical Engineers for

PRESSURE VESSELS AT THE ABOVE LOCATION AND FIELD SITES CONTROLLED BY THIS LOCATION

in accordance with the applicable rules of the Boiler and Pressure Vessel Code of The American Society of Mechanical Engineers. The use of the Code symbol and the authority granted by this certificate of authorization are subject to the provisions of the agreement set forth in the application. Any construction stamped with this symbol shall have been built strictly in accordance with the provisions of the Boiler and Pressure Vessel Code of The American Society of Mechanical Engineers.

THIS AUTHORIZATION *expires on* DECEMBER 3, 1984

Authorized on DECEMBER 3, 1981 *for*

THE AMERICAN SOCIETY OF MECHANICAL ENGINEERS
by the BOILER AND PRESSURE VESSEL COMMITTEE

Chairman Walter L. Harding

Secretary GM Eisenberg

Director, Accreditation Arlen A. Spadafino

Figure 4.1 *ASME certificate of authorization*

(d) certification (including use of authorized inspector).

2 Appointment of accredited manufacturers, authorized to use the 'U' symbol of the ASME. The manufacturers are under an obligation to comply with the requirements of the code and their competence is established by an initial management audit and subsequent review at regular intervals.

3 An obligation upon the manufacturer to make an inspection contract with an authorized inspection agency as defined in the ASME Code. The authorized inspector is a representative of this agency.

4 The code does not aim to cover every detail of the required design and construction. However, subject to the approval of the authorized inspector, the manufacturer has to detail his methods to make clear that they are as safe as those for which rules have been provided.

The specific management functions required of the manufacturing company by the representatives of the Boiler and Pressure Vessel Committee, who carry out the audit, include:

1 A clearly identified individual who has appropriate qualifications, authority and responsibility for all aspects of quality control needed to maintain the appropriate standards. In Fletcher & Stewart (FS), this individual was given the title of quality control manager (QCM).

2 A clear definition of the QCM's duties which included a direct responsibility for the maintenance of quality standards both in works manufacture and in field erection. These two mainline but self-contained activities of FS had been established under the control of the manufacturing and project directors respectively. The QCM was, however, made accountable to the technical director of FS, and as such had direct access to the highest management of the company. In so far as the ASME Code is concerned, the relevant FS procedures were incorporated into a manual, and in the event of disagreements or discrepancies which might not be resolved in accordance with the manual, the final decision would be taken by the technical director in strict compliance with the requirements of the code. The QCM would inform the technical director of the efficiency of the work by completion of a weekly scrap and reject report. The agreement of the authorized inspector would have to be obtained before any revision to the manual could be implemented.

3 Well-established procedures for the approval of design

calculations and drawings, and for their subsequent revision to provide clear verification of compliance with the code's requirements. Copies of original calculations, drawings and revisions to be made available to the authorized inspector on request.

4 Well-established procedures for the purchase, necessary inspection and custody of materials, so that certification meets the requirements of the code. Stock material not to be used for manufacture unless it has been re-requisitioned through the correct procedure.
5 Welding procedures, specifications, non-destructive testing and/or heat treatment requirements to be specified on the drawings. Proven procedures for the issue and control of drawings and specifications including the recall and change procedures.
6 Welding operators to be qualified by performance testing in accordance with the code and qualifications recorded in a welding operator's performance qualification test record. These have to be renewed at regular intervals and welding materials are subject to special procedures of purchase, storage, issue and use.
7 At each inspection phase appropriate to the manufacture or erection an inspector (accountable to the QCM) will record his findings by signing and dating a progress ticket. The authorized inspector will record his findings on the same ticket as necessary.
8 Non-destructive testing (NDT) personnel to be qualified and certified in accordance with the code, and the authorizing inspector may require requalification of NDT procedure and personnel at any time.
9 Certain forms in routine use, as detailed in the manual, can only be modified with the agreement of the authorized inspector. They include:
   (a) weekly scrap and reject report;
   (b) unit specification sheet;
   (c) commercial order form;
   (d) drawing issue record;
   (e) drawing office advance requisition;
   (f) assembly list sheet;
   (g) purchase order form;
   (h) purchase requisition;
   (i) stock replenishment order card;
   (j) goods received note;

(k) materials inspection report;
(l) rejection report;
(m) works progress sheet;
(n) material requisition form;
(o) deviation from drawing note;
(p) replacement demand note.

It is apparent from these brief extracts that the management disciplines imposed upon a manufacturing organization which works to the ASME Code are strict, but it is also evident that many of these management procedures and rules will find a parallel in the ways in which a chemical process plant should be governed when safety is at stake.

It should not be supposed that such procedures are fixed – they are constantly reviewed – or that they represent some ultimate set of standard practices. For example, there is a more stringent ASME code than the one described for common pressure vessels. This is known as ASME III, Division 1, for which the relevant stamp is 'N' and against which high-integrity nuclear pressure vessels are manufactured. Furthermore, in the case of the proposed pressurized water reactor for the Sizewell 'B' nuclear power station the CEGB has made extra special management arrangements. It is proposed that a management contract be confirmed with the Westinghouse Company of the USA for the supply of the complete system of equipment associated with the pressure vessel. This is because the original design of this system belongs to Westinghouse which has considerable worldwide experience of managing the supply and erection of such nuclear plant. However, the manufacture of the pressure vessel itself will be undertaken by Framatome of France, which is the only European company with experience of making this type of vessel. The largest forgings for this vessel are being bought from Japan as it is believed that this is the only country with a proven capability to manufacture such steel to the required material specification.

Westinghouse, Framatome and the Japanese steel company have their own arrangements for inspection. The CEGB also has its own in-house inspection services, but has placed a contract with Lloyd's to provide an independent assessment of all aspects of provision of the reactor pressure vessel. The CEGB has also placed contracts for inspection of the reactor pressure vessel forgings and welds with Babcock Power Ltd. All ultrasonic inspections of the vessel are further subject to validation by the Inspection Validation Centre (IVC), which is a wholly owned subsidiary of the UK

Atomic Energy Agency. IVC is overseen by an advisory committee whose members are drawn from universities and industries independent of the nuclear industry.

Thus, the CEGB's project management is at the centre of a web of many interacting contractual arrangements for the design and supply of this pressure vessel to a safety code. Although, pending the outcome of the Sizewell 'B' inquiry, official approval was not given to allow the CEGB to go ahead with the project as a whole, approval was given to place orders for the supply of the pressure vessel, due to the inevitably long delivery time.

A process management checklist

A technique commonly practised when undertaking a preliminary audit of a company's management competence is to run through a checklist of activities previously identified as likely to be relevant in the circumstances.

The checklist in Table 4.2 relates to the management organization of a chemical process manufacturing company viewed from the safety requirement. It is intended to be illustrative and introductory rather than comprehensive.

Table 4.2
A process management checklist

1 Is there a manual of safey codes endorsed by the top management, and in everyday use by the line?
2 Is there a named individual responsible for safety across the company?
3 Does he have active duties and access to top management?
4 Is there a procedure for resolving unforeseen disputes arising from the safety code?
5 Is operator training routine and are there tests of operator response to emergency?
6 Are there qualification standards and performance reviews of key individuals?
7 Is there a planned preventive maintenance programme to secure the safe performance of critical items?
8 Are there effective arrangements for the custody and retrieval of key operating and maintenance records?
9 Is there a system of safety permits and procedures for the control of on-line plant maintenance?
10 Are there regular emergency practices to test communications and a clear line of command?

If an audit were envisaged to check the practical application of such a checklist it would require for example:

1 production and study of the safety manual to compare it with those of other reputable companies;
2 perusal of the named individual's job description and related man specification, as well as an interview;
3 since it is inconceivable that an active safety manager would not find himself in occasional dispute with the line management, an audit would be concerned to see how lessons from a dispute were made clear to all concerned.

There are a number of management texts on the application of checklists to management practices and the use of objective criteria to validate performance, notably by John Humble and his associates.[6]

The application of more detailed checklists will be illustrated in Chapter 11, which discusses ways of mitigating and preventing hazards. The role of management also reappears in later chapters, but first the methodology of quantitative risk assessment must be examined in some detail.

References

1 Bradford, W. J., *What codes can be of help to practising engineers*, p. 57a, A.I.Ch.E. 19th Annual Loss Prevention Symposium, Houston, March 1985.
2 Gagliardi, D. V., *Dow's loss prevention principles*, p. 576, A.I.Ch.E. 19th Annual Loss Prevention Symposium, Houston, March 1985.
3 Carr, B., *Chlor-aid – the intercompany collaboration for chlorine engineers*, 1985 International Chlorine Symposium, London: Soc. Chem. Ind./Ellis Horwood, p. 113, 1986.
4 Davenport, J. A., *A study of vapour cloud incidents – an update*, I. Chem. E. Fourth International Symposium on Loss Prevention, Harrogate, September 1983. c1.
5 Withers, R. M. J., 'The trade and technology of sugar', Interdisciplinary Science Reviews, 6, p. 132, 1983.
6 Humble, J. W., *Improving Business Results*, New York: McGraw-Hill, 1967; *Management by Objectives in Action*, McGraw-Hill, 1970.

5 Quantifying the release – how big is a hole?

The quantitative estimation of all possible releases due to failures and unreliability in a large process plant is a truly formidable task. It may not even be possible to predict with certainty the magnitude of an individual release since there must be doubts about the possible size of the hole which may be formed by failure of the container wall through which the hazardous material would be released.

A relatively easy but superficial approach is to postulate the worst possible release, for example by assuming the total failure of a storage vessel and the immediate discharge of the entire contents into the surrounding atmosphere. While such a postulation may be credible in some cases, the likelihood may be so small that the risk from such an event becomes insignificant when compared to other risks. In other cases the likelihood may be more significant but, when taken in conjunction with the likelihood of the other events needed to complete the disaster scenario, the combined frequency may become so low as to be negligible.

By way of illustration, the case of ammonia storage, already mentioned in Chapters 2 and 3, may be cited. Ammonia is a toxic gas, manufactured and stored in large quantities over many locations, and can be lethal if high concentrations are inhaled as a result of an accidental release.

There are two modes of bulk storage, either in a vessel under pressure or in a refrigerated tank. The ammonia will be in liquid form, but in the former case the quantity will usually be limited to 500 tonnes, whereas in the latter storage tanks of 25 000 tonnes are in use. It is technically possible for a pressure vessel to fail catastrophically and the entire contents to be discharged directly into the air to form a cold and dense vapour cloud. A catastrophic rupture of a refrigerated tank so as to release the entire contents instantaneously is only conceivable in case of an earthquake. However, if an earthquake is possible the tank would be designed to take account of it, but possibly would not withstand the most severe earthquakes, which are of course unlikely in most regions of the world. Even in the event of a severe earthquake, liquid ammonia would not be ejected into the air but would overflow and probably be contained by a bund wall to some extent. The evaporation rate would depend on the weather conditions but, of course, the potential evaporation of 25 000 tonnes could form a very considerable hazard.

In the former case consideration must be given to the likelihood of pressure vessel failure, in the latter to the likelihood of earthquakes as well as to the structural strength of the tank. In both cases the likelihood of appropriate weather conditions would have to be taken into account when calculating the consequential dispersion patterns and their impact upon the surrounding locality.

There are many lesser, though much more likely, untoward events than the failure of storage vessels which constitute potential hazards. Interconnecting pipelines, valves, flanges and other appurtenances of the process are much more likely to fail, and although the quantities to be released are smaller they may still constitute a serious potential hazard and the risk may be much greater than from the 'top' event because of their higher failure rate. To obtain an estimate of the likely pattern of the lesser but more frequent events specific to a particular plant requires a careful survey. The methodology for this work needs to be appropriate to the need and to the resources available; for a large and complex process plant the task can be very onerous.

An essential first step is to identify all the possible serious hazards. Although many companies have used checklists for this purpose, industry has come to prefer the technique known as a 'hazard and operability study' or 'hazop'.[1]

'Hazop' and 'hazan'

It is usual to base the survey on a process schematic diagram showing the piping and instrumentation concepts (often called a P and I diagram). The production of P and I drawings using internationally agreed symbols has been standard practice in most organizations for many years, pioneered by such institutions as the Instrument Society of America. Nowadays these drawings are often prepared with the aid of a computer and automatic plotter. Such a diagram will have been prepared when the plant was first built, but it is important to check that it corresponds to the plant as it actually operates, since significant changes are often made in the light of working experience and the drawing may not have been updated.

A hazop study is best executed by a small team representing a wide spread of experience and knowledge, and a set of structured questions is presented to the first line of the P and I drawing leading to the first vessel. The set of questions is contained in a carefully chosen checklist of guide words, and provides a systematic search for all possible deviations of the system from normal. A checklist is illustrated in Table 5.1.

Table 5.1

A hazop guide word checklist

Guideword	Deviation
None	No forward flow or signal, or reverse flow
More of	Too high a flow or signal, pressure, temperature, etc.
Less of	Too low a flow, pressure, etc.
Part of	Wrong component composition
More than	Impurities present

The procedure is repeated for every other line entering the first vessel, and when all these lines have been examined the vessel itself is studied. Table 5.2 indicates how the guide word can initiate an action list. More detailed explanations appear in a booklet written by Trevor Kletz.[2]

Table 5.2

Hazop entry

Guideword	Deviation	Cause	Consequence	Action
Less of	Less flow	Leaking flange	Material loss	Regular inspection

The study can be readily extended to quantify the possible magnitude of the release but the frequency will have to be obtained by further analysis. A full quantitative examination involving both magnitude and frequency is often referred to as 'hazard analysis' or 'hazan'.

A hazan study may be able to provide an estimate of the frequency of a possible failure from past operating records, but if the fault has never occurred or only very rarely, the frequency will have to be synthesized perhaps by a full fault tree analysis, when there are several possible ways and a number of contributory factors are involved. Figure 5.1 provides a simple illustration of a fault tree. In such a case it will be necessary to use generic data of component failure frequencies from databanks provided by reliability studies. There are a number of sources of such data including relevant professional institutions such as the IEEE, public authorities such as the UKAEA, and manufacturers.

Note that in Figure 5.1 all the frequencies are quoted as annual failure rates. For the relief valve it would be more correct to use a probability number corresponding to the chance of failure on demand.

In Chapter 10 event trees are introduced. Whereas fault trees portray the relationship of the various faults which contribute to a 'top' event such as a vessel failure, an event tree starts with a single release and portrays all the possible events which might occur thereafter.

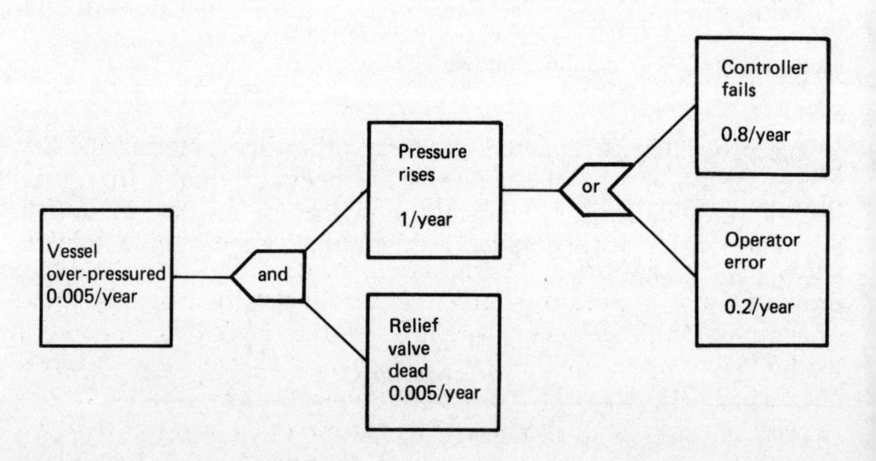

Figure 5.1 *A simple fault tree*

Two particular requirements in the application of hazan are worth emphasizing:

1. the large amount of effort needed in any on-site inspection, data collection, and calculation of predicted releases;
2. the need to relate any such prediction to historical records of similar events.

For example, Cox and Comer[3] have made clear that following a survey and calculation of the magnitude and the frequency of all possible releases due to failures, the many various failure cases should be reduced to a smaller number of 'equivalent discrete failures' before the quantitative assessment is undertaken, and that the associated failure frequencies be based on generic classified data. The latter is obtained from historical records.

At the present time there is some research activity attempting to develop and apply 'expert system' computer methods directly to the P and I data to obtain the results of a hazan examination much more rapidly and it is hoped more securely.[4] Of course, there can be no substitute for direct contact with the plant operation, since P and I drawings do not provide information on the relative location and siting of equipment. These can be of vital importance to safety.

Hazop and hazan studies most often take place long after the major design parameters are frozen. Indeed, they may well take place after the plant has been built and is in operation. Thus, if hazards are identified by such studies they can only be controlled by adding on protective equipment and alarm systems, but it is not usually possible to remove the hazard by changes in design.

The execution of a project is a phased process and at the stage when outline planning permission for a new installation is requested the design is usually only an outline appropriate to the feasibility phase of the project master plan. Nevertheless, such an outline plan should be sufficient for the purposes of a risk assessment, although it is clearly inadequate for a hazop or hazan study. A hazan study team would ideally comprise members whose experience is drawn from a variety of sources. It would benefit from the views of commissioning engineers as well as process designers. In addition to the information obtained from P and I diagrams, the team will need plot plans, piping and pumping specifications and drawings as well as generic reliability data from a range of items. Those should be on-site inspections as well as written communication.

While a risk assessment based wholly upon generic data will be sufficient when seeking outline planning permission, it seems probable that if there is a Public Inquiry a fuller assessment will be demanded. There will normally be an appreciable period between a planning application and an Inquiry and this could give the company time to carry through a preliminary hazop and hazan study and to update its safety case as appropriate.

Because of the effort involved and the time that may be consumed in a full hazan study, there is a need for a simple method of release quantification so that an approximate calculation can be formulated in an event tree of all the relationships which make up the total risk estimate. It may well be found that, on the basis of such a first approximation, no further analysis is necessary, as an accident is either very unlikely or will have such trivial consequences that no more money should be spent on it. On the other hand, the first approximation may indicate a potentially serious situation which warrants further investigation.

One possible solution to this need is provided by the use of 'standard' release data. Based on an analysis of historical frequency and magnitude data taken from collections of case histories, it has been suggested[5] that a standardized release pattern may be used which has the form

$$\log_{10}T = a \log_{10}f + b$$

where T is the magnitude of the release in tonnes, f is the frequency of the release in events per annum divided by 10 000, and a and b are parameters.

It is generally the case that only the reports of the more serious accidental releases have been recorded in the literature, and it is necessary to consider the extent of this under-reporting before any attempt is made to establish a 'standard' release pattern such as might be represented by this equation.

Compensation for under-reporting

An interesting paper by Badoux[6] discusses various ways in which data sets which suffer from under-reporting may be corrected. It gives the theoretical background to the logarithmic transformations which enable a view to be taken of the likely distribution of events in time with special reference to the Pareto distribution.

Badoux illustrates this application of the Pareto distribution to the study of reported accidents by two examples taken from the petrochemicals industry, the second of which, due to Wiekema, relates to historical data on vapour cloud incidents.

Wiekema[7] gathered data on 162 vapour cloud incidents in the USA over the years 1932–81. Sixty-two of these were ignitions and provided sufficient information on the amounts of materials released. Pareto analysis was applied and it was found, by applying a linear regression analysis to a logarithmic transformation of all releases exceeding 10 tonnes, that the compensated total number of releases (for the full set of 162) approximated 8000. In this analysis it was assumed that the original data set was complete for all sizes greater than 10 tonnes. Extrapolation of the straight line obtained from the linear regression so as to intercept the Y-axis gives a simply derived estimate for $\ln N$ and hence the total number of incidents. Figures 5.2 and 5.3 reproduced from Badoux's paper make the procedure clear.

Badoux claims that the total of 8000 derived from the restricted set of 62 is realistic and represents an average number of 160 per annum. If it had been assumed that the original data set was complete for all releases greater than 2 tonnes, the analysis would

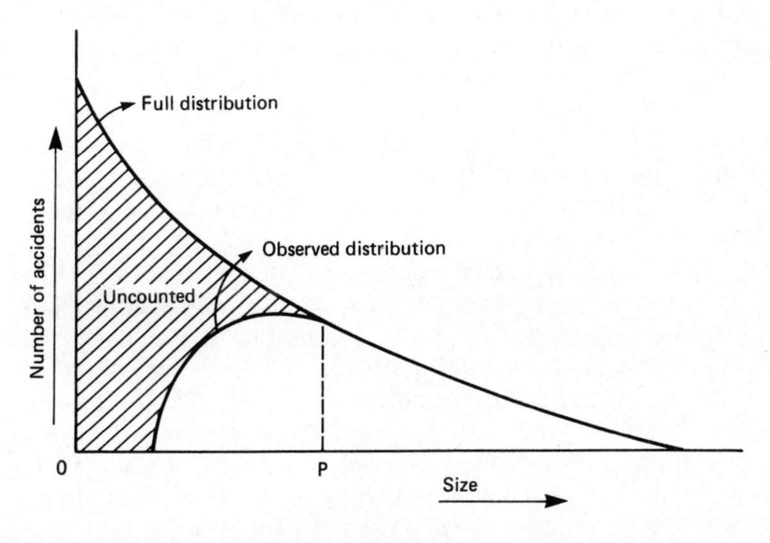

Figure 5.2 *Full and observed distribution for the size of an accident*

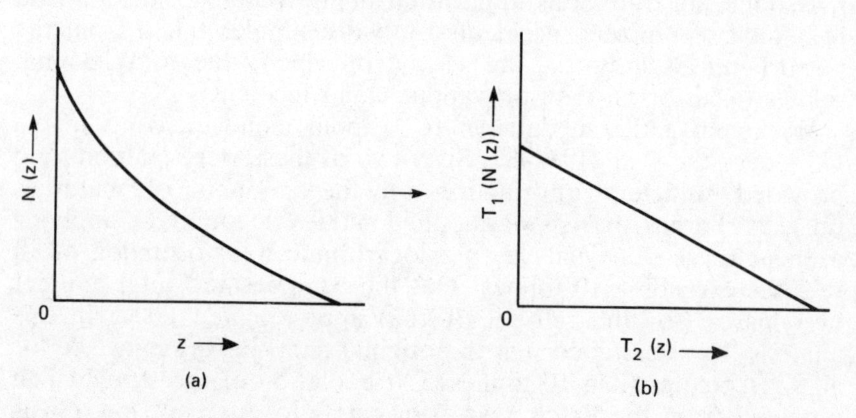

Figure 5.3 *Transformation of N(z) and of z*

have predicted a complete set of approximately 3000. It is obvious that the method may introduce a measure of conservatism into the predicted outcome depending on the view taken of the degree of completeness in the original set.

The Davenport list

Probably the best known list of vapour cloud explosions is that collated by Davenport[8] and most recently published at the Institution of Chemical Engineers' Loss Prevention Symposium in 1983. From this list of 71 incidents, 30 have been selected, where both release size and deaths have been reported. The time span for the 30 incidents is thirty-seven years and the number of events for various categories over this period provide the data for a frequency/magnitude plot. Table 5.3 shows the initial data for both the fatalities and the release size. The data is also shown in Figure 5.4. From the line of best fit through the points an asymptotic straight line gives the augmented cumulative frequency line from which the finite frequency numbers in the last column are derived. It will be apparent that the original data set seemed to be complete at sizes above 5 tonnes and is therefore consistent with Badoux's interpretation of Wiekema's data set of similar releases. Examination of Figure 5.4 suggests that there is a 3:1 ratio between tonnes released and the number of fatalities. This may seem rather a close ratio but it must be remembered that a selection has been made from Davenport's list to exclude incidents where no deaths took place. The finite frequency data gives the equations:

$$\log 500 = a \log .002 + b$$

$$\log 2 = a \log .075 + b$$

from which:

$$a = -1.52$$

$$b = -1.41$$

The Kletz list

This list, compiled by Kletz and associates[9] at ICI, spans the years 1970–81. It covers a whole range of worldwide incidents in the process industries and a ten-year total of 778 fires and explosions involving 1196 fatalities has been extracted. Unfortunately, it is acknowledged that the completeness of the reporting diminishes as the distance from the UK increases, and while there is frequent enumeration of deaths and injuries, the size of the release is not given so often. It has therefore proved necessary to select from the list those incidents reporting fatalities, and then to transform the fatality magnitude into a release magnitude using the simple 3:1 relationship obtained from Figure 5.4.

While it is only the slope of the relationship between the frequency and magnitude which is of interest, for purposes of comparison it is convenient to adjust the absolute magnitude of the sets to represent a common constituency.

From the Kletz list the total number of deaths worldwide from major accidents involving five or more fatalities is around 1500, the number in the Netherlands is 22, in the UK 34 and in the USA 206. The ratios between these numbers are in line with corresponding data on general accidents in the chemicals and process industry. In the Rijnmond Report it is stated that the Dutch authorities envisage that there are 1000 notifiable plant installations in the Netherlands likely to cause a major hazard. In consequence, to facilitate a comparative display of the slopes of the Davenport and Kletz data sets with other sets, their magnitudes have been divided by 70 000 to make them notionally representative of a single-plant constituency. The results are detailed in Table 5.4 and Figure 5.5.

Table 5.3

Frequency/magnitude data based upon Davenport

Fatality magnitude	Actual cumulative frequency	Release magnitude	Actual cumulative frequency	Best fit cumulative frequency	Augmented cumulative frequency	Augmented finite frequency
200	$.004 \times 10^{-4}$	500	$.004 \times 10^{-4}$	.003	.003	.002
100	$.004 \times 10^{-4}$	200	$.007 \times 10^{-4}$	.006	.006	.003
50	$.007 \times 10^{-4}$	100	$.007 \times 10^{-4}$	.010	.010	.004
20	$.011 \times 10^{-4}$	50	$.020 \times 10^{-4}$	.017	.017	.007
10	$.027 \times 10^{-4}$	20	$.039 \times 10^{-4}$	.033	.033	.016
5	$.045 \times 10^{-4}$	10	$.057 \times 10^{-4}$	.054	.054	.021
2	$.091 \times 10^{-4}$	5	$.081 \times 10^{-4}$	.078	.093	.039
1	$.115 \times 10^{-4}$	2	$.088 \times 10^{-4}$	.097	.168	.075
		1	$.105 \times 10^{-4}$	.102	.300	.132
		0.5	$.111 \times 10^{-4}$	.112	.528	—
		0.2	$.116 \times 10^{-4}$	.116	—	—

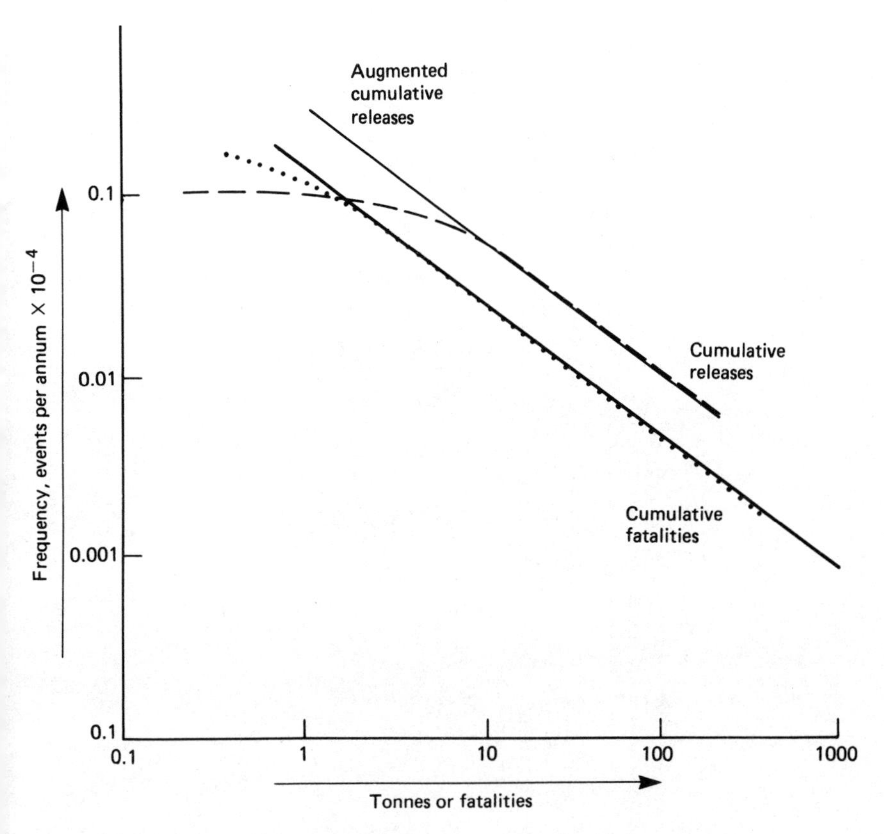

Figure 5.4 *Frequency/magnitude plots of Davenport data*

The finite frequency data gives the equations:

$$\log 300 = a \log .025 + b$$

$$\log 3 = a \log 2.0 + b$$

from which

$$a = -1.05$$

$$b = +0.79$$

The Fawcett list

This list together with supplements is contained in a report which is internal to the UK Ministry of Works,[10] and provides the

Table 5.4
Frequency/magnitude data based upon Kletz

Fatality magnitude	Actual cumulative frequency	Best fit cumulative frequency	Augmented cumulative frequency	Augmented finite frequency	Release magnitude (sc.2)	Release magnitude (sc.3)
100	$.03 \times 10^{-4}$	$.03 \times 10^{-4}$	.035	.025	300	2000
50	$.10 \times 10^{-4}$	$.09 \times 10^{-4}$	.09	.055	150	1000
20	$.20 \times 10^{-4}$	$.20 \times 10^{-4}$	.20	.11	60	500
10	$.50 \times 10^{-4}$	$.50 \times 10^{-4}$	.50	.30	30	200
5	1.05×10^{-4}	$.85 \times 10^{-4}$	.85	.35	15	100
4	1.3×10^{-4}	1.3×10^{-4}	1.30	.45	12	80
3	1.7×10^{-4}	1.85×10^{-4}	1.90	.60	9	60
2	2.3×10^{-4}	2.3×10^{-4}	3.00	1.10	6	40
1	3.3×10^{-4}	3.4×10^{-4}	5.00	2	3	20

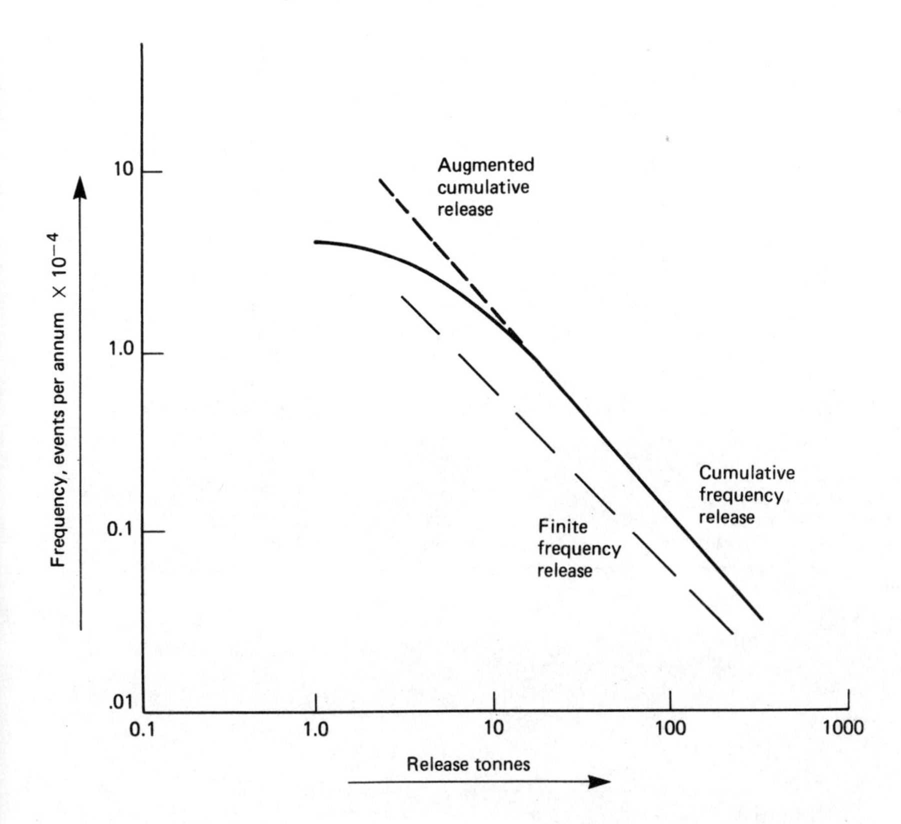

Figure 5.5 *Frequency/magnitude plots of Kletz data*

background data for a paper given by Jarrett to the New York Academy of Sciences[11] in 1968. Fawcett's list of 74 accidental explosions ranging in size from 500 lb to 168 000 lb up to 1938 was supplemented in the report by 24 further explosions during World War 2 at munitions ships and stores ranging in size from 300 lb to over 5 million lb.

The analysis of this data in Table 5.5 follows the two previous patterns with the addition of an estimated frequency for explosions in peacetime. This was obtained by comparing the number of incidents in 1914–45 with those during 1918–38.

For purposes of comparison, and following the explanation given in the previous section, the frequency data has been divided by a factor of 1000.

Table 5.5
Frequency/magnitude data based upon Fawcett

Release magnitude (tons)	Actual cumulative frequency (p.a./10 000)	Augmented cumulative frequency (p.a./10 000)	Augmented finite frequency (p.a./10 000)	Estimated peacetime finite frequency (p.a./10 000)
2 000	.140	.190	.080	.004
1 000	.280	.280	.090	.004
500	.430	.420	.140	.007
200	.710	.700	.280	.014
100	.860	1.000	.300	.015
50	1.140	1.600	.600	.030
20	1.280	2.600	1.000	.050
10	2.570	4.000	1.400	.070
5	3.000	5.700	1.700	.085
2	3.860	9.600	3.900	.195
1	6.140	14.000	4.400	.220

The finite frequency data gives the following equations:

$$\log 2000 = a \log .080 + b$$

$$\log 2 = a \log 3.90 + b$$

from which

$$a = -1.76$$

$$b = +1.36$$

Cremer and Warner studies

It is of interest to compare the three previous examples from historic records of actual events with the predictions for single installations contained in risk assessments that have been published. Well-known and authoritative risk assessments by Cremer and Warner appear in their Rijnmond report[12] – they relate to six individual installations but only three of them appear here, namely:

1 the Oxirane propylene plant at Seinehaven;
2 the AKZO chlorine plant at Botlek;
3 the UKF ammonia plant at Pernis.

The tonnages released and their associated frequencies for the three installations may be found in the tables provided by Cremer and Warner in the relevant sections of their Rijnmond Report. They are based for the most part upon site surveys, fault tree systems analyses, checklists, and generic failure rate data. The data and augmented frequencies are reproduced in Tables 5.6, 5.7 and 5.8.

For propylene, the finite frequency data gives the equations:

$$\log 1000 = a \log .0035 + b$$

$$\log 1 = a \log 2.1 + b$$

from which

$$a = -1.08$$

$$b = +0.35$$

Table 5.6

Frequency/magnitude data for propylene (Rijnmond)

Release (tonnes)	Actual finite calculated frequency $\times 10^{-4}$	Best fit cumulative frequency	Augmented cumulative frequency	Augmented finite frequency
1 000	—	.0055	.0055	.0035
600	.00359			
500	—	.011	.011	.006
300	.0023			
200	—	.030	.03	.019
104	.24	.050	.05	.20
100				
50	—	.100		.050
25	.0035			
20	—	.250	.250	.150
15	.011			
10	—	.500	.500	.250
7	.42			
5	—	.700	.900	.400
2	—	1.00	2.2	1.300
1.5	.42			
1.0	—	1.200	4.3	2.10
.5	—	1.350	8.0	3.70
.2	—	1.50	2.0	12.00
.1	.08	1.6		

Table 5.7
Frequency/magnitude data for chlorine (Rijnmond)

Release (tonnes)	Actual finite calculated frequency × 10^{-4}	Derived best fit cumulative frequency	Augmented cumulative frequency	Augmented finite frequency
100	.00093	.003	.003	.002
50	.00740	.008	.008	.005
50	.0170		—	
20		.065	.065	.057
10		.130	.240	.175
5		.300	0.90	0.660
4.5	.0660			
2.5	.2600			
2.3	.0160		—	
2		.600	3.50	2.60
1	1.200	1.1	10.00	6.50

This compares with the average slope of -1.44 for the historical data sets which relate to releases involving fire and explosion.

The predictions for the toxic releases have a steeper slope.

For chlorine, the finite frequency data gives the equations:

$$\log 100 = a \log .002 + b$$

$$\log 2 = a \log 2.6 + b$$

from which

$$a = -0.55$$

$$b = +0.53$$

For ammonia, the finite frequency data gives the equations:

$$\log 500 = a \log .0005 + b$$

$$\log 2 = a \log 3.9 + b$$

from which

$$a = -0.62$$

$$b = +0.66$$

Table 5.8
Frequency/magnitude data for ammonia (Rijnmond)

Release (tonnes)	Actual finite calculated frequency $\times 10^{-4}$	Derived best fit cumulative frequency	Augmented cumulative frequency	Augmented finite frequency
682	.00023	.00023		.0005
500		.0005	.0005	
250	.0018			.0015
200	.0005	.0021	.002	.005
100		.0075	.007	.013
50		.0200	.020	.080
20		.070	.100	.250
10		.150	.350	
9.2	.0044			
7	1.9			
6.6	.014			
5		.400	1.100	.750
2		1.000	5.00	3.900
1.9	.0044			

The average 'Rijnmond' slope is -0.75 whilst the average of the historical data sets is -1.44, giving a mean of -1.10.

Figure 5.6 provides a graphic comparison of all six slopes.

As a first approximation to a typical release pattern a standard value of $a = -1.0$ is suggested. Alternative values of -1.3 and -0.7 may be used to check the sensitivity of the outcome to changes in the slope. Before any attempt can be made to apply this relationship it is necessary to discuss the frequency which may be given to the top event, as this provides a direct basis for assigning a value to b. This is not an easy matter.

Top event frequency estimation

The data given in Tables 5.3 to 5.8 and displayed in Figure 5.6 suggests that the frequency of the top event may be expected to be in the range 2 to 200 times in 100 million years. Such top events are frequently expected to be the consequence of the catastrophic failure of a large containment vessel, leading to the nearly instantaneous discharge of its total contents. Catastrophic vessel failures have been the subject of a number of studies. For example,

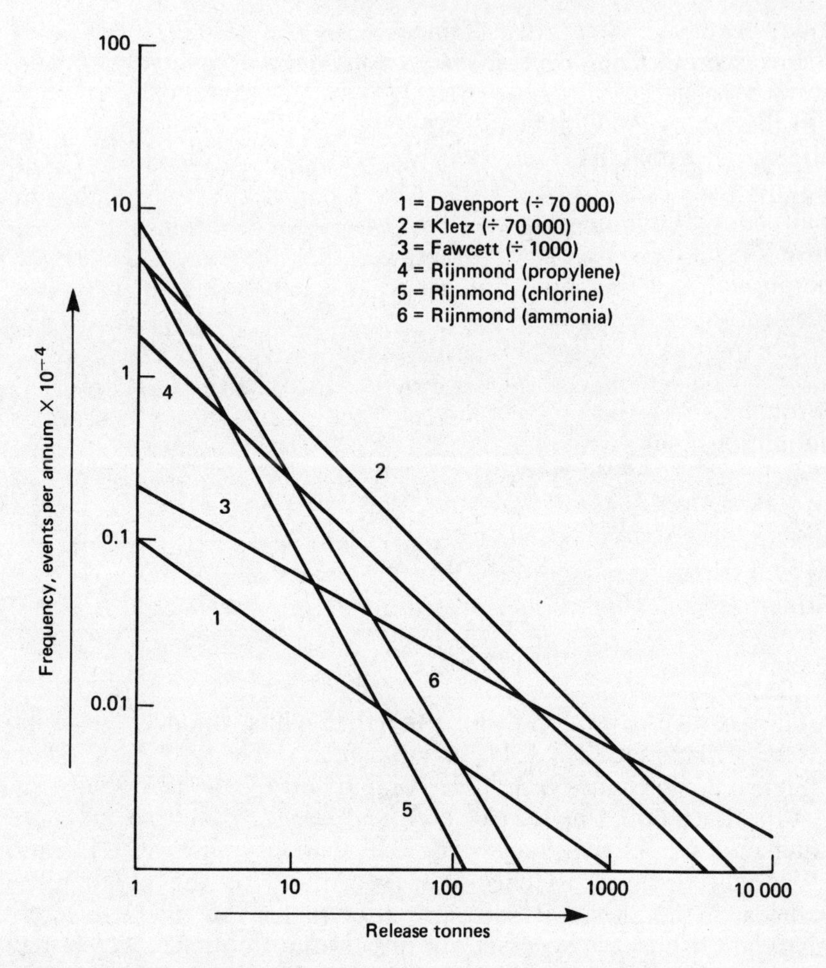

Figure 5.6 *Comparison of the six 'slopes'*

in the UK Smith and Warwick of SRD (Safety and Reliability Directorate of the United Kingdom Atomic Energy Authority) have published a survey[13] and there have been rather more comprehensive surveys in the USA[14] and West Germany.[15]

From these historical statistics it is possible to make a probabilistic assessment of the likely annual failure rate, and in 1975 Bush suggested that there is a 99 per cent confidence at the upper bound for less than one failure in 100 000 vessel years when built to ASME

Code, Section 1. Since this figure is based on historical data and lessons from working experience have been applied to present and future constructions, it is conservative for new installations.

In the most demanding cases there are further doubts about the validity of probabilistic assessments. In addition to the inherent weakness of the statistical evidence of the rare events based on small populations, it is often impossible to distinguish the real cause of failure. For example, the vessel may have failed for operational reasons and not from any inherent defect in its specification or construction.

An exhaustive debate on such issues took place at the Sizewell inquiry, where evidence on the results of a historical survey gave a probabilistic failure 'upper bound' rate in the region of two in one million reactor years. The CEGB objected strongly to the use of such an estimate as the basis for the top event failure rate and advocated instead a figure derived from a deterministic mathematical model based on stress analysis. This gave a figure 100 times lower than the probabilistic 'upper bound'.[16]

Independent evidence presented at the inquiry by local authorities and others supported the deterministic approach provided it was allied to strong management measures in quality assurance through all the procurement phases. This requires much engineering time and effort but is nevertheless necessary for installations which might otherwise constitute a major hazard.

References

1 Kletz, T. A., *Hazop and Hazan*, Rugby: I. Chem. E., 1983.
2 Ibid.
3 Cox, R. A. and Comer, P. J., 'Development of low cost risk analysis methods for process plant', *I. Chem. E. Symposium Series 71*, pp. 14–16, April 1982.
4 Poucet, A., *CAFTS, Computer Aided Fault Tree Analysis*, ANS/-NS Int. Top. Meet. on Prob. Safety Methods and Applications, p. 115, San Francisco, February 1985.
5 Withers, R. M. J., *Foundations for single computer models*, Loughborough University of Technology MHC/86/2, pp. 4–8, 1986.
6 Badoux, R. A., *Some experiences of a consulting statistician in industrial safety*, Proceedings of the Fourth Nat. Rel. Conference 3B, 1983.

7 Wiekema, B. J., *Analysis of vapour cloud accidents*, Proceedings of the Fourth Euredata Conference, Venice, 1983. c1.

8 Davenport, J. A., *A study of vapour cloud incidents – an update*, I. Chem. E. Fourth International Symposium on Loss Prevention, Harrogate, September 1983.

9 Kletz, T. A. and Turner, E., *Is the number of serious accidents increasing?*, ICI Safety Note 79/2B, London: Chem. Ind. Assn., 1979.

10 'Notes on the basis of outside safety distances for explosives involving the risk of mass explosion', ESTC, 3/7/Explos./43, 1959.

11 Jarrett, D., 'Derivation of the British Explosives Safety Distances', *Annals of the New York Academy of Sciences*, 152(1), pp. 18–35, 1968.

12 *Rijnmond Report*, Dordrecht, Netherlands: Reidel, 1982.

13 Smith, A. and Warwick, *Second Survey of Defects in Pressure Vessels*, SRD R30, London: HMSO, 1974.

14 Bush, S., 'Pressure vessel reliability', *Trans. ASME*, pp. 54–70, February 1975.

15 Kellerman, O., 'Unfallanalyse in der Kerntechnik', *Technische Überwachungs* 13, Nr. 11, November 1982.

16 Sizewell 'B' Inquiry: Days 216, 217, 230 and 243; Supporting NII/S/92, LPA/P/4 (Add. 2), GLC/P/6 plus Adds.

6 Quantifying dispersion – how long is a piece of string?

In the previous chapter, the concept of a standardized release pattern to provide the primary input to a risk estimate was explained. This can provide the foundation for a simple estimating procedure, which uses as its prime independent variable a release pattern made up of discrete masses together with their associated release frequencies.

This chapter explains how in these simplified estimates every escape or release can be considered equivalent to a discrete mass and then relates the dispersion of such a mass to the next appropriate dependent variable in the chain of events which has to be worked through.

The dependent variable that has been found most appropriate is the downwind range to a given gas concentration. This in turn may relate to a lethality criterion in the case of a toxic gas or to a flammability criterion in the case of a gas which may ignite. Such criteria of toxicity and flammability and their simplified algebraic relationships to the damage which may be inflicted upon an impacted population are explained in Chapters 9 and 10.

The selection of the downwind range as the primary dependent variable is partly justified by the number of experimental observations and theoretical predictions which have appeared in the literature. This has made it possible to determine a scaling law which relates the downwind range to the mass released. It has also been justified by the subsequent use of the scaling law in computer-based models which have demonstrated reasonable agreements with data from case histories and more rigorous models.

An account is given of the effect upon the scaling law of secondary independent variables such as the weather and the physical properties of the gas.

The concept of equivalent mass

It is convenient to divide the manner in which hazardous liquids and vapours may be released into two classes: the near instantaneous release from a vessel which has suffered catastrophic failure; and the slower release from a partial failure of the vessel or from a pipe or other appurtenance of the vessel system. The first class of event is readily related to an instantaneous discrete mass release while the second has to be described in the first instance in terms of a mass flow.

While the second class is much the more likely to occur in practice, it is the first class which provides the largest instantaneous release. Since the development of a standard release pattern may rest upon an estimation of the sizes and frequencies of the larger possible events, it is often helpful to transform mass flows into equivalent instantaneous mass releases when constructing and comparing release patterns with generic and plant-specific data.

In the case of instantaneous release from a pressure vessel one can simply state that most if not all of the vessel contents are released. The thermodynamic equation for the flash-off of a superheated liquid from a pressurized container is well established and readily found in standard texts.[1]

$$\varnothing = 1 - \exp\left[-\frac{Cp}{\Delta Hv}(\theta_1 - \theta_2)\right] \tag{1}$$

where $\varnothing$ = mass fraction vapourized;
Cp = specific heat of the liquid;
ΔHv = latent heat of the liquid;
θ_1 = storage temperature;
θ_2 = boiling point of stored liquid.

In the case of escape from a catastrophically ruptured container, however, turbulence caused by the rapid boil will add spray to the flash fraction given by calculation from the formula (1). This may result in an ejection of the total contents of the vessel as a spray of liquid droplets. The considerable turbulence will quickly result in the entrainment of air into a cloud formation, perhaps by as much as a factor of ten.

For example, in the case of chlorine storage in the UK formula (1) suggests the flash fraction would be around 20 per cent, but it may be assumed that the spray will increase this to 45 per cent, while the mass of the final cloud with its entrained air will be 250 per cent of the original contents.

Mathematical models of the dispersion of such dense clouds therefore begin with assumptions on the initial shape of the cloud which are somewhat conjectural. The proposed simple scaling law, described later, by-passes this difficulty.

Where the release is not instantaneous, three distinct kinds of release may be considered:

1 non-flashing flow;
2 flashing liquid flow;
3 gaseous discharge.

The first of these is relatively straightforward as Bernoulli's equation applies:

$$\dot{m} = Ka\sqrt{[2\rho(P_1 - P_2)]} \qquad (2)$$

where
$\dot{m}$ = mass discharge rate;
K = discharge coefficient;
a = area of discharge;
ρ = liquid density;
P_1 = upstream pressure;
P_2 = downstream pressure.

But the determination of flashing flow from fundamental theory is beset by a number of problems. However, a simple approximate method has been described in the second Canvey report[2] and empirical methods have been described elsewhere.[3]

Whilst empirical methods may not be applicable in all cases, it is often found in multiphase flow that the reduction factor between each of the three phases is 4. This simple approximation gives for multiphase flow an addition of 25 per cent to the non-flashing flow for flashing flow, and 6 per cent for sonic gas release.

By way of illustration the multiphase release of butane from a broken 50 mm pipe has been calculated. Application of formula

(2) gives the non-flashing flow as 17.5 kg per second, if the upstream pressure is 200 KPa at −5°C. The approximate flashing flow is 4.3 kg per second and the sonic release 1.0 kg per second making a total rate of 23 kg per second.

To transpose such mass release rates into equivalent cloud masses the method outlined by Marshall is followed.[4] He has derived the following equations for the equivalent mass of material in a cloud between flammable limits:

Atmospheric dispersion:

$$Q_{FL} = \left(\frac{0.321}{D^{0.59}}\right)\left(\frac{\dot{m}_0^{1.59}}{\bar{u}^{1.59}}\right)\left(\frac{1}{\chi_L^{0.59}} - \frac{1}{\chi_U^{0.59}}\right) \tag{3}$$

Jet dispersion:

$$Q_{FL} = \left(3.40\rho_a^{1.5}\right)\left(\frac{\dot{m}_0^{1.5}}{w_0^{1.5}}\right)\left(\frac{1}{\chi_L^2} - \frac{1}{\chi_U^2}\right) \tag{4}$$

where Q_{FL} = quantity in the cloud (kg);
D = a constant;
$\dot{m}$ = mass flow rate (kg per second);
u = wind velocity (metres per second);
χ_L = concentration at lower limit (kg per metre);
χ_U = concentration at upper limit (kg per metre);
ρ_a = density of air (kg per metre);
w = jet velocity (metres per second).

These relationships are illustrated in Figure 6.1, taken from Marshall's paper, which refers to a hypothetical hydrocarbon release with lower and upper flammability limits of 0.039 and 0.176 respectively.

The diagrams in Figure 6.1 suggest the following approximations:

At u = 10 metres per second neutral (D) weather:

$$Q_{FL} \simeq 0.65m^{1.59} \tag{5}$$

At u = 1 metre per second stable (F) weather:

$$Q_{FL} \simeq 60m^{1.59} \tag{6}$$

The diagrams also suggest that even with very high leak rates the cloud will comprise up to 50 tonnes only, between flammable

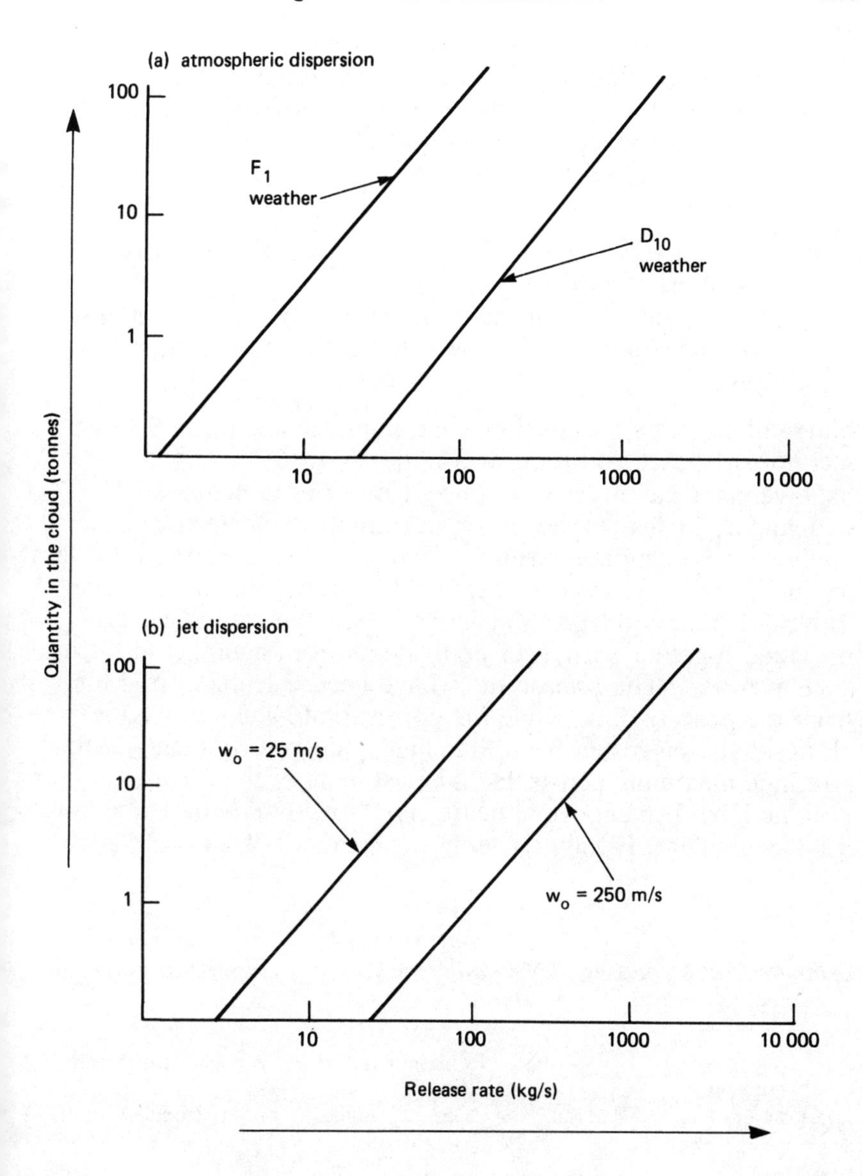

Figure 6.1 *Marshall's equivalent mass relationships*

limits under the worst conditions. It has been remarked also, that for hydrocarbon clouds, the amount between flammable limits is 20 per cent of the total and these constraints provide scope for considerable simplification at the outset of a risk assessment.

Low-mass rate releases are unlikely to impact the general population off site since:

1. the relatively low release rates gives small equivalent masses and short hazard ranges;
2. many gases only form denser-than-air clouds under catastrophic conditions. At low release rates they drift harmlessly upwards.

Marshall suggests an absolute minimum release rate of 10 kg per second for the constitution of a major hazard.

Davenport has utilized a concept of equivalent mass, the TNT equivalent, in his survey of vapour cloud explosions.[5] This TNT equivalent is computed from a survey of the damage and relates to the estimated mass of explosive causing similar damage. In Table 6.1, adapted from Marshall's paper, typical release rates are provided together with TNT equivalents from actual case histories by Davenport. The release rates have been calculated by Marshall from the process data, while the corresponding cloud mass is given alongside as suggested from Marshall's diagrams but subject to the 50-tonne maximum constraint. The estimate of the release quantity obtained by Davenport appears next together with Davenport's TNT equivalent. Finally Davenport's estimate of the yield is given.

Table 6.1

Estimated rates, masses, TNT equivalents, and yields for actual vapour cloud explosions

Location	Material	Release rate (kg)	Cloud size (tonnes)	Mass (tonnes)	Equivalent TNT (tonnes)	Yield (%)
Port Hudson	C H	23	7	55	45	7.5
Beek	C H	36	10	5	1	4
Pernis	Hydrocarbon	100	50	78	20	6
L. Charles	i-C H	100	50	9	11	10
E-St Louis	C H	400	50	53	2	2
Decatur	i-C H	660	50	69	20–125	3–18
Flixborough	C H	1030	50	36	18	5

This is the ratio of the TNT equivalent and the energy content of the release quantity given by Davenport.

The level of agreement between the various estimates is typical and in line with informed expectations.

Neutral density modelling

The theory of the dispersion of gases which have a density close to that of air goes back over many years.[6] It has a practical application to the calculation of smoke dispersion. It can also be applied to the dispersion of dense gases if their concentration is low, and has some conceptual value in the development of a theory for heavy gas dispersion.

It is usual to express the variation in concentration in a plume under steady-state conditions as a function of wind speed, distance and time, using rectangular co-ordinates in three dimensions and using the concept of a diffusion coefficient. The development of various equations by a number of workers has been reviewed by Lees.[7] Of particular interest is the prediction of the ground level concentration downwind along the axis of the plume, and this is shown to be proportional to the reciprocal of the distance and directly to the magnitude of the release.

Experimental work on the relation between concentration and distance has been described by Sutton in 1953 who demonstrated that the downwind concentration on the ground along the plume centre line varied in proportion to $x^{-1.76}$, where x is the downwind distance.

Sutton enhanced the original theory to bring the predicted results more into line with practical observation. His work was further extended by Pasquill in 1961, who placed the stability of the atmosphere into six categories A–F in terms of the time of year, of the day and the night-time,[8] and the presence or absence of cloud. Category A is very unstable, F is very stable, D being neutral. A seventh category, G, is sometimes added. In England, category D occurs more than 50 per cent of the time for winds in excess of 7 metres per second, but as the wind drops below 7 metres per second the other categories become more important. Categories E and F can only occur at night, categories A and B can only occur during the daytime.

In the final version of his formulae he gave a relationship for the ground-level concentration on the axis as follows:

$$C = \frac{2.8 \times 10^{-3} Q}{u\,d\,h\,\theta} \tag{7}$$

where C = downwind concentration (gm per metre);
d = downwind range (km);
h = vertical spread (m);
Q = mass rate of release (gm per minute);
u = wind speed (metres per second);
θ = lateral spread (degrees);

which suggests that, for a continuous release, the downwind concentration is inversely proportional to the distance and directly proportional to the mass.

Table 6.2 gives the variation in θ according to Pasquill.

The value of h varies from around 300 at 100 km to around 6 at 1 km in stable/neutral conditions, according to the following formulae:

(D) $\log h = 1.85 + 0.835 \log d - 0.010(\log d)^2$

(F) $\log h = 1.48 + 0.656 \log d - 0.122(\log d)^2$

Table 6.2
Pasquill stability and lateral spread

Category		Frequency	θ, d = 100 m	θ, d = 100 km
A	unstable conditions	22%	60	20
B			45	20
C			30	10
D	neutral conditions	65%	20	10
E	stable conditions	13%	15	5
F			10	5

The ground-level concentration at the centre of a neutral density cloud or puff-type release is given by a similar type of equation but in this case use is made of three dispersion coefficients in the three directions (downwind, crosswind and vertical).

$$C = \frac{2Qi}{(2\pi)^{3/2}\partial_x \partial_y \partial_z} \tag{8}$$

This also represents the maximum concentration on the axis at a particular point. The total integrated dose at this point may be needed for predicting the effects of toxic gas, and this is given by:

$$D = \frac{Qi}{\pi \, \partial_y \partial_z u} \qquad (9)$$

where u = wind speed.

The values of $\partial_y \partial_z \partial_x$ are obtained by interpolation from the data in Table 6.3. This is a somewhat shaky procedure much dependent upon a satisfactory understanding of the weather.

Table 6.3

Neutral density diffusion coefficients

	x = 100 m		x = 4 km	
Pasquill category	$\partial_x = \partial_y$	∂_z	$\partial_x = \partial_y$	∂_z
Unstable (A–C)	10	15	300	220
Neutral (D)	4	3.8	120	50
Stable (E–F)	1.3	0.8	35	7

Whilst the original theory has been modified by the incorporation of empirical data, the prediction of neutral density gas dispersion seems to be well established, therefore, but subject to two severe constraints:

1 the difficulty of understanding the weather and predicting the wind speed;
2 the nature of turbulent flow involves inherent variation in the dispersion so that repeats are unlikely even though an average result may be obtained from the results of repeated releases.

There is some uncertainty about the relevance of the Pasquill categories to the atmospheric turbulence and diffusion factors affecting the dispersion of dense gas clouds.[9] This uncertainty is one of the limiting factors to the application of deterministic mathematical models of dense gas dispersion based on fundamental laws of physics. Attempts have been made to measure such factors directly by scientific instruments but a wide scatter has been reported in the percentage occurrence of the various properties and categories depending on whichever observation protocol has been adopted.

Wind speed is capable of precise definition and measurement and in many countries tables are published which give the relative time the wind may blow into given sectors, at various speeds, at separate locations. In England the wind does not blow about 11 per cent of the time, blows up to 7 metres per second for 30 per cent of the time, and in excess of that speed for about 60 per cent of the time. The light winds around 2 metres per second provide the conditions for maximum downwind range, but these do not normally exhibit a strong prevailing direction. The stronger winds do exhibit a prevailing direction from the south-west.

Dense gas dispersion models

Dense gas dispersion models have to take into account three distinct phases of gas behaviour:

1 initial mixing (source models);
2 gravity slumping (dense gas models);
3 turbulent spread (neutral gas models).

The background to the first of these has been discussed in the introduction to this chapter. The chief areas of doubt and uncertainty are:

1 the quantity of liquid droplets forced out of the ruptured vessel by turbulence;
2 the quantity of air entrained during the initial release;

and they critically determine whether the initial gas cloud is substantially denser than air.

The basic concept of gravity slumping involves the balancing of the downward force due to the density difference between the gas cloud and air against an opposite force associated with the disturbed kinetic energy of the air. Application of Bernoulli's theorem gives:

$$\frac{\partial R}{\partial k} = k\sqrt{\frac{\Delta\rho}{\rho_a} g H} \qquad (10)$$

where R = radius of the cloud;
k = some arbitrary constant;
$\Delta\rho$ = difference between the two densities;
ρ_a = density of air;
g = gravitational constant;
H = cloud height.

Strict proof from the theorem gives a value of $k = 2$, but practical observation suggests a value nearer to 1.

The third phase of turbulent spread is usually modelled on the precepts of Pasquill, previously described.

There are two main categories of mathematical models used to represent dense gas dispersion:

1 'box' models;
2 'conservation' models.

The box models[10,11] are based upon the concept of the initial cloud slumping as a whole and incorporate a formula which is based upon equation (10). It gives a rate of advance of the cloud height which is proportional to a simple function of cloud height and density, further advanced by the action of the wind.

Some models terminate the slumping phase without considering further air entrainment or edge mixing, while others do take this into account.

Any dense cloud that is colder than ambient will absorb heat which will affect the density. Due to the flatness of the cloud the effect can be considerable, altering the entrainment rate and bringing on the transition to neutral density conditions much sooner.

The various models attempt to quantify these heat transfer effects in differing ways. There are also a number of criteria in use for establishing the changeover point from the phase of gravity slumping to that of turbulent spread.

The differing choices that are possible in these models lead to differing predictions of cloud growth and in the early years these differences were quite startling, particularly for the larger sizes of hazard for which it is almost impossible to contemplate any experimental verification. One of the first reviews of this situation, by Havens,[12] showed that the predicted consequences of a spillage of 25 000 cubic metres of LNG on to water varied from 0.75 miles to 50 miles in so far as its possible ignition was concerned. There have since been a number of improvements in the model-building, and the differences have narrowed appreciably. They will be compared in a later section of this chapter.

Quite apart from the fact that the 'box' models do not come up with unanimous results, particularly for the larger releases, they have further disadvantages:

1 they do not allow for the effects of buildings and topography;
2 they cannot provide any estimate of the relationship between peak and mean concentrations;

3 they cannot take into account any time-varying nature of the source.

Nevertheless, they have proved extremely useful in helping to provide a systematic basis for the examination of observations from actual incidents, for the planning of experiments and trials with real gases, for the comparison of collected data, and for the extrapolation of the small-scale test data to hypothetical large-scale releases.

In an effort to make further progress there has been much development in recent years of mathematical models based on fundamental laws of conservation of such properties as mass, momentum and energy.[13] A typical 'conservation' model will comprise six basic non-linear partial differential equations, and these have to be solved by digital computer programs involving finite element techniques. Such solutions have tended to be rather complex numerically and expensive in computer time. However, this is a rapidly changing field of activity, and a variety of computer packages is becoming available which can be executed on personal computers and the results displayed as three-dimensional plots using conventional dot matrix printers.

Simultaneously, there have been a number of substantial field trials, backed up by experimental work using wind tunnels. There is a fair amount of common ground and much of it has been reviewed in the literature. Table 6.4 provides a comparative summary of the various model outcomes.

However, the predictions are still strongly affected by somewhat speculative weather dependent diffusivity coefficients, and there remains the problem that the concentrations are random quantities, and averaged results are not necessarily sufficiently descriptive of the problem. Moreover, these mathematical models do take up much computer time if they are incorporated into programs which involve many branches on an event tree and multiple sources. For such work a simple scaling law has many advantages, and such a scaling law is described in the next section.

A downwind scaling law

In the absence of wind or other effects it is obvious that the simplest relationship between the cloud mass and the distance from the cloud centre to a given concentration must be of the form:

Table 6.4

Various model outcomes

Name	Model type Description
Germeles & Drake	Well mixed cylindrical source, slumping transition to Gaussian plume no edge mixing
DENZ	Well mixed cylindrical source, slumping transition to Gaussian plume no edge mixing
Cox & Carpenter	Well mixed cylindrical source, transition with heat transfer and edge mixing
Shell HEDAGAZ	Needs a wind but no transition needed, no heat transfer but edge mixing
EIDSVIK	Well mixed source, no transition needed, with edge mixing
Mariah Zephyr SIAB FEM SIGMET	Conservation models, fully three-dimensional, no transition needed. All these models use virtually the same basic equations but have differing numerical solution methods with differing approximations in matters of detail.

Some comparative results

Predictions for down wind range to LFL of LNG

Model	Source weather and wind (metres per second)	10000T F2	10000T D5	1000T D5	100T D5	12T C5	14T D8	10T E2	8T D6
G & D		22.2	5.8	2.5	1.2	.13	.15	.66	.24
DENZ		5.6	4.0	1.6	0.6			.35	
C & C		5.0	5.0	1.8	0.8			.50	
Shell		9.9	3.0	1.3	0.5			.50	
SIAB						.22	.26	.42	.32
FEM						.19	.21	.63	.38
SIGMET		8.8	4.0	2.5	1.4	.15	.18	.60	.22

$$\text{Range} = \text{constant} \times (\text{Mass})^{\frac{1}{3}}$$

In the second Canvey report published by the HSC an empirical equation for the radius of a vapour cloud has been given:

$$R = 30 \times M^{\frac{1}{3}}$$

It is based on an assumption that the ratio of the cloud radius to its height is 5:1. According to Lees,[14] this agrees well with the observed shape of clouds during the incidents at Flixborough and Beek.

However, for a neutral-density dispersion we have the well-founded Pasquill equation (7) given previously, which suggests that for a continuous release, the downwind concentration is directly proportional to the mass.

It is reasonable to suppose that for a puff release the result will lie somewhere between these two, and this is confirmed by evidence from the Warren Springs wind tunnel as given in Table 6.5 and Figure 6.2.

Hence, for an average release:

$$R = (1/C)^{0.76} \tag{11}$$

Table 6.5
Warren Springs wind tunnel results

Conc. %	Log C. %	R,d2×a	R,d4×a	1/R2	Log 1/R2
30	1.477	20m	50m	.05 m^{-1}	−1.381
20	1.301	30m	60m	.033	−1.481
10	1.000	50m	60m	.020	−1.699
5	0.699	88m	110m	.012	−1.921
4	0.602	100m	135m	.018	−2.000
3	0.477	120m	165m	.008	−2.097
2	0.301	170m	220m	.006	−2.222
1	0.0	320m		.003	−2.523

Intercept = − 2.47
Slope = 0.762
Cor. coef. = 0.9975

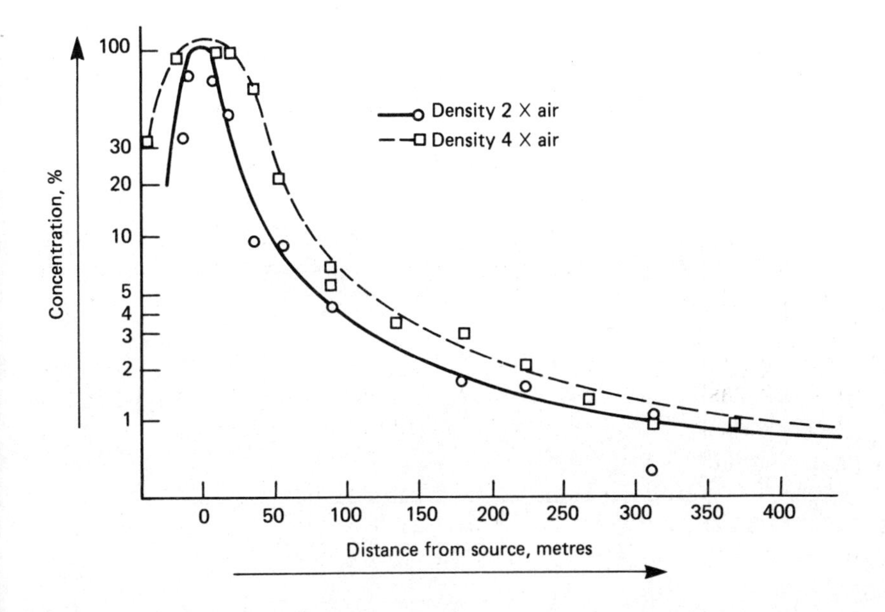

Figure 6.2 *Warren Springs wind tunnel results*

whilst, since over the range of concentrations 1–5 per cent there is an increase of range by 35 per cent for double the released mass, giving:

$$R = \mathrm{k} \times (\mathrm{Mass})^{0.43}$$

A similar result has been suggested in the second Canvey report, based upon the DENZ computer model. This gave an approximate downwind range over the release range 10–1000 tonnes according to the expression:

$$R = \mathrm{k} \times (\mathrm{Mass})^{0.4}$$

but the dependence of range upon density is likely to be a complex affair and what may be true for concentrations down to 2 per cent may not apply to the levels below 0.1 per cent which relate to hazards from toxic gases. Figure 6.3 taken from the Warren Springs report[15] shows somewhat hypothetically the effect of decreasing

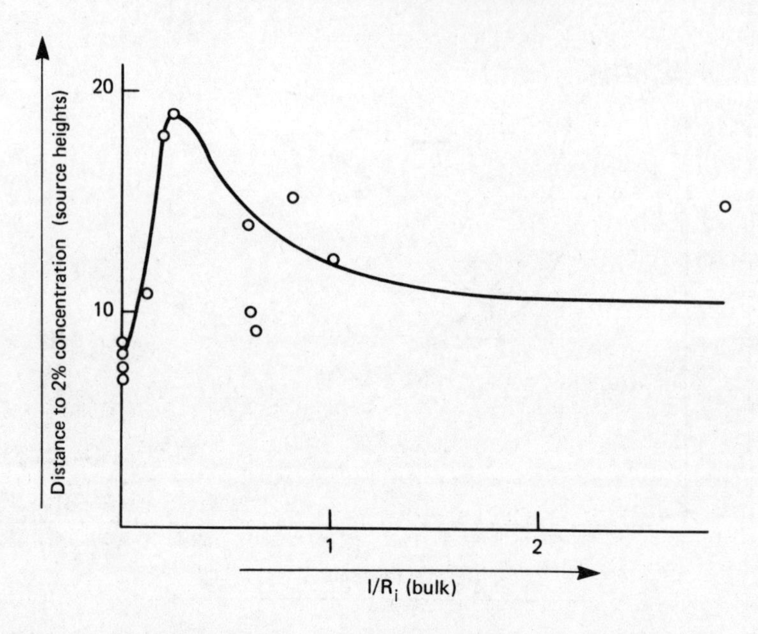

Figure 6.3 *Bulk Richardson number versus downwind range*

the bulk Richardson number upon the downwind range to 2 per cent concentration.

The Richardson number is given by the following expression:

$$R = g\frac{\Delta\rho}{\rho_a}\cdot\frac{L}{u^2} \tag{12}$$

where g = acceleration due to gravity;
$\Delta\rho = \rho_g - \rho_a$ = difference of gas density to air;
L = length scale;
u = wind speed.

Light winds and heavy gases allow the cloud to drift downwind without undue dispersion. Initial increase of wind speed or decrease of density increases the range, but as the winds become stronger and the gas gets lighter the dispersion rate increases so that the range decreases again.

The Warren Springs wind tunnel work[16] gave results which agreed with full-scale trials at Porton Down,[17] and also provided a satisfactory prediction for the field trials at Thorney Island. But the results of these trials, involving a released mass of 0.007 and 0.36 tonnes respectively, require cautious extrapolation to the large releases envisaged in major hazards.

There are only a limited number of other field observations with which to compare or enhance the above work. It is necessary to restrict the data to that available for the one gas LNG and to correct for small differences in wind speed using Figure 6.3. This has provided a set of ten results over the release range 0.007–26.0 tonnes, and they are summarized in Table 6.6.

The range, adjusted for D5 conditions, is given to the lower limit of flammability, except for E4 and E6. These (from the SS *Gadilla* trials) are also anomalous in that the release was for ten minutes' duration and not instantaneous. The recorded range was a visible observation of cloud size believed to be much longer than the LFL because of the high ambient humidity. In Table 6.6 the range for E4 is predicted by regression line from the set A–D and E6 is obtained by ratio from the visual observations of E4/E6.

The details of the regression analysis are:

1 Correlation coefficient = 0.968.
2 Intercept (at log M = 0) = −1.03.
3 Slope = 0.369.

These results give too narrow a base for extrapolation to the tonnages envisaged in major hazards and so the regression analysis has been extended by the addition of ten further results selected from four different computer simulations; DENZ, Cox and Carpenter, Shell Hedagaz and SIGMET. These models, which have been comparatively reviewed in the literature,[18] were briefly contrasted in Table 6.4.

Table 6.7 gives details of the further regression analysis and a summary of the findings. The average value of 0.42 given to the slope from the set of regression analyses provides a value for the index in a simple scaling law. It clearly gives a conservative result in comparison with the results from the field trials, but as is shown in Table 6.7 there is not much practical difference at sizes below 50 tonnes. It has previously been suggested that this is likely to be the maximum cloud size, and if interactive computer models are used for estimating the risk it is an easy matter to change the value of this index if desired.

Table 6.6
Field trial test results

Name	Description	Results: Test no.	Tonnes	Range at D5 (km)
Shell *Gadilla* (1973)	6 tests 10–70 tonnes of LNG discharged on sea	E 4	26	.398
		E 6	67	.554
Shell Maplin (1980)	34 propane and LNG tests 2–7 tonnes on water	D 15	6.3	.128
		D 39	3.6	.122
		D 56	1.3	.107
'Burro' China (1980)	8 tests 14 tonnes LNG over water then sand	C 8	9.9	.291
		C 7	13.8	.196
HSE Porton (1977)	35 tests 40 m of Freon over grassland	B 33	.007	.019
HSE Thorney (1982)	15 tests 2000 m Freon over grassland	A 7	.36	.068
		A 8	.36	.047

Table 6.7
Field tests plus models regression analysis

Correlation coefficient = 0.973
Intercept (at log M = 0) = −1.01
Slope = 0.41

Comparative summary of findings

Mass released (tonnes)	Field test regression (km)	Field plus models regression (km)	SIGMET model (km)	DENZ model (km)
.01	.017	.014		
.1	.040	.037		
1.0	.093	.097		
10	.218	.250	.24	
100	.510	.647	1.4	.60
1000	1.191	1.67	2.5	1.6
10000	2.786	4.32	4.0	4.0
'slope'	'0.37'	'0.41'	'0.44'	'0.4'

Thus the derived scaling law is:

$$R = a \times M^{0.42} \qquad (13)$$

where R is in kilometres;
M is in tonnes.

The factor a depends on the physical properties of the dense gas and the weather conditions, as will be explained later. But it must be emphasized that the scaling law merely provides an estimate of the mean consequence from a number of releases. In practice the difference between a particular release and this average may be considerable.

This point is illustrated by Figure 6.4 adapted from the Warren Springs report. It shows the repeatability of the concentration/time relationship at a point equivalent to 10 metres, downwind from the source at a wind speed of 5 metres per second. Over twenty repeat runs the smallest peak concentration was 4 per cent and the largest was 36 per cent, with a mean of 15 per cent. The gas density was twice that of air. This variation seemed to result from turbulence within the cloud rather than from major changes in cloud shape or direction of travel.

It must also be emphasized that the derivation of the relationship is essentially pragmatic. Its chief justification is that it provides a simple basis in a computer model for calculating the downwind

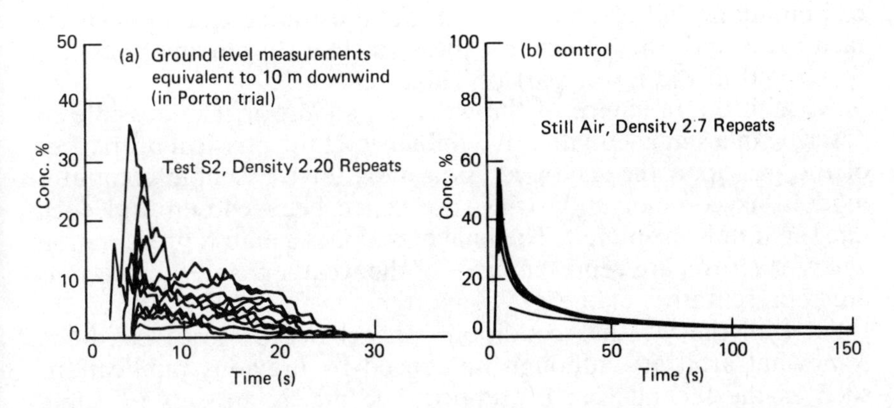

Figure 6.4 *Variability in repeat runs in Warren Springs wind tunnel*

range. It gives results which are reasonably compatible with the limited range of the field trials and to the results from more elaborate calculations. It is difficult to assess the relative merits of the alternative solutions at the present time and since the proposed relationship steers a middle course between them all, it is hoped that it will be found more acceptable than most.

Values for the factor *a*

Two important aspects of the weather which affect dispersion are, first, the speed and direction of the wind and, second, the stability of the atmosphere.

The results obtained from risk estimates which take these considerations into account show an overwhelming individual risk during the night as compared to the daytime in so far as the risks from drifting vapour clouds are concerned. This is due to the Pasquill categories E and F only occurring at night. Moreover, these night-time individual risk contours do not show much asymmetry, and graphical displays are most conveniently shown as a set of concentric circles. The risk contours for the daytime conditions cover a much smaller area and are elliptical with the long axis reflecting the prevailing wind direction, but for most practical purposes the combined average day and night contours may be represented by circles.

Table 6.8, which gives the probability of the weather at Canvey, England and at Zestienhoven, Netherlands, illustrates some of the previous remarks. Only five categories of weather and wind combinations will be considered in the illustrative examples of the factor a, and the effect of these further approximations are illustrated in the lower part of Table 6.8.

As with the influence of the weather, so also is it impossible to describe in a single equation the influence of the physical properties of the gas upon the factor a. As a basis for the simple computer models six common industrial gases have been chosen and each one separately examined. The final outcome is a matrix presentation where the rows are representative of the weathers and the columns are representative of the gas properties.

As with the weather conditions, the choice of gases has been somewhat arbitrary although influenced by previous publications such as the second Canvey report. The matrix appears in Table 6.9, which also lists the key physical data leading to the calculated

Table 6.8
Weather probabilities

Place	Weather A	B	C	D	E	F	All weathers
Canvey	.003	.045	.035	.812	.049	.046	1.00
Zestienhoven							
Night	0	0	—	.631	—	.369	1.00
Day	—	.246	—	.753	0	0	1.00
Night and day	—	.123	—	.692	—	.184	1.00

Place	Weather C5	D5	D2	E2	F2	All weathers
Canvey	.022	.267	.303	.035	.034	1.00
Zestienhoven	.061	.251	.111	.120	.113	1.00

values of *a*. The explanation of the derivation of these *a* values for each gas is given subsequently starting with the base case of LNG at the D5 weather and wind condition.

LNG

A value of $a = 0.095$ at the D5 condition fits the data given in the previous section when the index value of 0.42 is given to the scaling formula.

From equation (12) and Figure 6.3, which suggests a maximum range when the Richardson number is 4, and using the density values given in Table 6.9, the maximum range for the peak concentration of LNG occurs at a wind velocity of 2.3 metres per second. Figure 6.3 suggests a value for *a* of 0.11 at the D2 condition, relative to .094 at D5.

Since the effect of the weather condition is somewhat conjectural the mean findings from three published model results have been used. Thus, in a paper by Havens,[19] results from the SIGMET model are given which show an increase in the downwind range by a factor of 1.66 for LNG as the condition changes from D to F. In a paper by Megson and Griffiths,[20] results from the DENZ model are given which suggest a corresponding factor of 1.25, whilst in the Rijnmond Report[21] the Cox and Carpenter model

Table 6.9

Initial conditions for gas dispersion scaling law

Property	Gas					
	Ammonia	Chlorine	Propane	LNG	Butane	Hydrogen fluoride
Gas density	0.68	2.49	1.50	0.55	2.00	0.99
Boiling point, deg. C	−33	−35	−45	−161	−2.0	+19
Liquid S. G.	0.82	1.56	0.58	0.35	0.65	1.42
Flash fraction	0.15	0.18	0.28	0.58	0.10	0.03
Vapour weight (kg)	m	m	m	m	0.8m	m
Added air	9.6m	4m	11.2m	13.0m	2.8m	5.2m
Cloud weight	10.6m	5m	12.2m	14.0m	3.6m	6.2m
Cloud kg/cu.m	1.68	1.71	1.41	1.52	1.64	1.33
Cloud deg. C.	−50	−45	−16	−66	−45	−7
Cloud cu. m.	6.63	2.92	8.65	9.21	2.19	4.68
LFL/LC_{50} at 30 m	1.0%	.05%	2.1%	5.8%	1.8%	0.12%
u for R maximum	2.3 m/s	2.5 m/s	1.9 m/s	2.3 m/s	2.4 m/s	1.6 m/s
a for C5	0.10	0.20	0.10	0.07	0.09	0.10
a for D5	0.14	0.28	0.15	0.09	0.12	0.13
a for D2	0.17	0.26	0.20	0.11	0.11	0.15
a for E2	0.20	0.31	0.24	0.13	0.13	0.18
a for F2	0.23	0.36	0.29	0.15	0.15	0.21

m = numerical value of weight of gas contained in cloud, in kg demonstrates a factor of 1.12. The average value of these findings is 1.34 and is used throughout Table 6.9; it gives a value for LNG at F2 of 0.15. For E2 an intermediate value of 0.13 applies, and for C5 0.07.

Propane

Flash fraction and air entrainment calculations, based on the equations given previously and summarized in Table 6.9, suggest that the final cloud masses for propane and LNG will be in the ratio 12.2/14.0. This ratio is then raised to the power 0.42 to obtain a factor which can convert the D5 LNG figure to propane. But the LFL densities are in the ratio 5:2.1 and this ratio is raised to the power 0.76 from equation (11) to obtain a further conversion factor.

From equation (12) and Figure 6.3 the maximum range occurs at a speed of 1.9 metres per second and, following the previous procedure and taking into account the conversion factors, a D5 value for *a* for propane becomes 0.15.

The only field test which provides a direct comparison between propane and LNG is that at Maplin, described by Puttock *et al.*[22] Comparison of spill no. 47 with no. 39 suggests a factor of 2.04 which would suggest a value for *a* for propane of 0.19.

In the second Canvey report the ratio between propane and LNG, based on the DENZ model, is given as 1.2. This suggests a value for *a* for propane of 0.11.

The average of these three results is 0.15, and this value is entered into Table 6.9. A similar procedure to that outlined above for LNG gives values for C5, D2, E2 and F2 of 0.1, 0.2, 0.24 and 0.29 respectively.

Ammonia

The calculations for flash fraction and air entrainment, summarized in Table 6.9, give a factor of 10.6/14.0 to be raised to the power 0.42 to convert from LNG to ammonia. The next conversion step requires an understanding of the toxicity level to be applied rather than the lower limit of flammability. In the cases of the Canvey and Rijnmond assessments a gas concentration of 3200 ppm was used. This figure was thought to be the concentration lethal to 50 per cent of a population exposed for 30 minutes. More recent work suggests that this level should be raised to 11 500 ppm, or 1.15 per cent. Following the procedures previously outlined gives a figure of 0.23 for *a* for ammonia at D5.

There is no direct field data with which to compare either LNG or propane with ammonia. There are a few case histories of ammonia spills, and the fragmentary evidence from the cases at Blair, Mcpherson, Potchetroom and Southwest Highway indicates that a value for *a* might be around 0.15.

The DENZ result reported by Megson and Griffiths and the Cox and Carpenter result given in the Rijnmond Report require modification to bring them into line with the revised toxicity figure of 11 500 ppm. Figures for *a* of 0.10 and 0.08 respectively are thus obtained.

The average of all four results is 0.14 and this is entered into Table 6.9. Following a similar procedure to that outlined for LNG gives values for *a* at C5, D2, E2 and F2 of 0.10, 0.17, 0.20 and 0.23 respectively.

Chlorine

The lethal concentration of chlorine is very much less than that of ammonia, and recent work has suggested an LC_{50} of 500 ppm at 30 minutes' exposure time. However, in the Rijnmond Report very much lower levels (approximately 50 ppm) were used, and there is much debate about the allowance to be made for activity rate, vulnerable people and shelter. The relationship given in equation (11), based on the Warren Springs wind tunnel work, is not necessarily valid at levels below one per cent. Thus, for example, the DENZ results for chlorine quoted by Megson and Griffiths provide a value for the index in equation (11) of 0.58, while the corresponding value for ammonia is 0.40. Following the previous procedures but using ammonia as the base case gives a D5 value for *a* for chlorine of 0.45, using a concentration value of 500 ppm. This is significantly larger than the results quoted by Megson and Griffiths or from the Rijnmond (Cox and Carpenter model) Report and, taking into account the density correction to 500 ppm, these values are 0.15 and 0.25 respectively.

It is not possible to obtain any meaningful quantitative information from case histories, although recent work on the records from World War 1[23] has given some support to the LC_{50} of 500 ppm. The value entered in Table 6.9 is 0.28, the average of the three.

From equation (12) and Figure 6.3, the wind speed for maximum range is 2.5 metres per second and the D2 value for *a* is 0.26. The values for C5, E2 and F2 are then 0.20, 0.31 and 0.36 respectively.

Butane

The flash fraction and air entrainment calculations for butane suggest that the initial cloud masses for butane and LNG are in the ratio 3.6:14.0. This ratio has then to be raised to the power 0.42 to obtain the required conversion factor. The LFL densities are the ratio 5:1.8 and this is raised to the power 0.76 (from equation (11)).

From equation (12) and Figure 6.3, the maximum range occurs at a wind speed of 2.4 metres per second. Following the procedures previously outlined gives *a* values at C5, D5, D2, E2 and F2 of 0.09, 0.12, 0.11, 0.13 and 0.15 respectively.

Hydrogen fluoride

The LC_{50} of hydrogen fluoride is subject to some uncertainty, but recent study of the literature suggests a value of 1200 ppm at 30 minutes' exposure time. Although in a similar concentration class to chlorine, it is roughly half as toxic and the results which follow are significantly less in range than has been suggested by the DENZ results quoted in the second Canvey report.

The molecular weight of hydrogen fluoride can vary due to the formation of dimer and hexamer, and commercially it may be found mixed with benzene or isobutane. However, simple flash fraction and air entrainment calculations suggest that the initial cloud masses for hydrogen fluoride and chlorine may be in the ratio 6.2:5 (it is assumed that the boiling point is a mixture of hydrogen fluoride and isobutane). The LC_{50}s are in the ratio 500:1200. Following the previous procedures leads to a wind speed for maximum range of 1.6 metres per second and using the previously derived D5 value for chlorine as a reference gives the values for *a* at C5, D5, D2, E2 and F2 of 0.10, 0.13, 0.15, 0.18 and 0.21 respectively.

Table 6.10
Some downwind ranges for chlorine and ammonia

Calculation	Concentration	Range(km)	
Ammonia (100-tonne release)		*at D5*	*at F2*
Scaling law LC_{50} at 30	10000 ppm	1.03	1.7
Scaling law LC_{50} at 10	17000 ppm	0.69	1.1
Denz, Canvey 2	3255 ppm	1.3	2.4
Denz, COVO(b)	8000 ppm	0.9	1.8
Cox and Carpenter, COVO(a)	3200 ppm	0.96	1.11
Cox and Carpenter, COVO(b)	8000 ppm	0.66	0.77
Chlorine (50-tonne release)			
Scaling law LC_{50} at 30	500 ppm	1.45	1.86
Scaling law LC_{50} at 10	866 ppm	1.10	1.41
Denz, USCGVM	48 ppm	2.6	6.5
Denz, COVO(b)	400 ppm	0.7	1.5
Cox and Carpenter, USCGVM	48 ppm	4.9	4.5
Cox and Carpenter, COVO(b)	400 ppm	1.3	1.2

Summary and Conclusion

In recent years the published estimates of the downwind range of a hazardous gas dispersion have tended to shorten. The chief reason for this is a better understanding of the dispersion process, reinforced in the case of the toxic gases chlorine and ammonia, by a raising of the lethal concentration levels.

Table 6.10 compares various downwind values of chlorine and ammonia for 50- and 100-tonne releases respectively. In addition to scaling law results with LC_{50} values at 10 and 30 minutes' exposure times, results are given from earlier calculations published in the literature, usually with more conservative toxicity levels. It must be emphasized, however, that all these results are mean values and that the scatter around the mean values is large. While the scatter is appreciable and is indicative of the uncertainty which still obtains, there is a fair measure of common ground and sufficient agreement for many aspects of hazard assessment.

References

1 Lees, F. L., *Loss Prevention in the Process Industries*, p. 426, London: Butterworths, 1980.

2 *Canvey Second Report*, London: HMSO, 1981.
3 Wallis, G. B., *Int. J. Multiphase Flow*, 6, p. 97, 1980; Akker, H. E., Snoey and Spoelstra, 'Discharges of pressurised liquefied gases through apertures and pipes', *I. Chem. E. Symposium Series 80*, 1, E23, 1983.
4 Marshall, J. G., *I. Chem. E. Symposium Series 58*, 11, 1980.
5 Davenport, J. A., *A study of vapour cloud incidents – an update*, I. Chem. E. Fourth International Symposium on Loss Prevention, Harrogate, September 1983. c1.
6 Taylor, G. I., 'Eddy motion in the atmosphere', *Phil. Trans. A*, 215, 1915.
7 Lees, op. cit. note 1, p. 439.
8 Pasquill, F., *Atmospheric Diffusion*, Chichester: Ellis Horwood, 1974.
9 Sedefian, L. and Bennett, A., 'Comparison of turbulence classification schemes', *Atmospheric Environment*, 14, p. 741, 1980.
10 Cox, R. and Carpenter, *Cloud dispersion model for hazard analysis*, Proceedings of Symposium on Heavy Gas, Frankfurt: Reidel, 1979.
11 Fryer, L. and Kaiser, *DENZ – A computer program for dispersion*, SRD/UKAEA R152, July 1979.
12 Havens, J. A., 'An assessment of the predictability of LNG vapour dispersion', *Journal of Hazardous Materials*, 3, p. 257, 1980.
13 Jagger, S. F., *The dispersion of dense, toxic, and flammable clouds in the atmosphere*, I. Mech. E. Seminar, September 1982.
14 Lees, op. cit. note 1, p. 439.
15 Hall, D. J. *et al.*, Warren Springs Laboratory Report LR394, 1982.
16 Ibid.
17 Picknett, R. G., 'Dispersion of dense gas puffs', *Atmospheric Environment*, 15, p. 509, 1981.
18 Havens, J. A., 'A description and assessment of the SIGMET LNG vapour dispersion model', *Chem. Eng. Mon.*, 16, p. 181, 1982, also Woodward, J. L., Havens, J. A. *et al.*, 'A comparison of several models of heavy vapour clouds', ibid., p. 161.
19 Havens, ibid.
20 Griffiths, R. and Megson, 'Effects of uncertainties on hazard range for ammonia and chlorine', *Atmos. Envir.*, 18, p. 1195, 1984.

21 *Rijnmond Report*, Dordrecht, Netherlands: Reidel, 1982.
22 Puttock, J. S., Blackmore and Colenbrander, 'Field experiments on dense gas dispersion', *Journal of Hazardous Materials*, 6, pp. 13–41, 1982.
23 Withers, R. M. J. and Lees, F. P., 'The assessment of major hazards: the lethal toxicity of chlorine: Part 3, Crosschecks from gas warfare', *Journal of Hazardous Materials*, 15, 1987.

7 The chances of fire and explosion

The two previous chapters provided a basis for specifying a pattern of release and the subsequent dispersion of material in the form of a heavy gas cloud.

In the case of inflammable material it is next necessary to provide an estimate of the chances of the material igniting. This may occur on site near to the point of release or it may occur off site as the cloud drifts away; it may not occur at all. If the material does ignite, the ignition may be followed by an explosion. Such an explosion may occur on site close to the ignition, off site or not at all.

Since the primary independent variable is the mass released, it is convenient to relate the chances of ignition and explosion to this quantity.

Reference has already been made to the work of Wiekema (Chapter 5) in the context of compensation for under-reporting.[1] Wiekema lists 162 events in the USA involving fire and explosion of which 62 can be classified in terms of size and whether they exploded or only ignited. In a presumed reference to Badoux's work,[2] Wiekema gives a total of 8000 events when the 162 are compensated for under-reporting. In the main this compensation enhances the numbers of small sizes of release. It implies that the full set corresponding to the 62 is 3040.

Table 7.1 gives the numbers in Wiekema's distribution. Those totalling 62 and 36 are supplied in Wiekema's paper, but those in the hypothetical set of 3040 are inferred.

Table 7.1
Ignition and explosion data after Wiekema

Size range (tonnes)	Total	Ignition	Ignition and explosion
>100	12	8	3
100–10	97	19	12
10–1	421	25	15
1–0.1	670	9	6
>0.1	1 840	1	0
Total	3 040	62	36

There are in the literature a limited number of estimates which provide a relationship between the mass released and the chances of ignition and explosion.[3] However, if these are combined with the Wiekema data, tabulated and smoothed by taking a linear regression through a log–log plot, the somewhat speculative pattern in Table 7.2 emerges for on-site conditions.

The chances of ignition and explosion should the cloud drift off site have been estimated in the Canvey report.[4] Should an unignited cloud reach the edge of a populated area, it is suggested that there will be a 70 per cent chance of ignition occurring at the edge, a 20 per cent chance at the centre and a 10 per cent chance of non-ignition. If the cloud is smaller than 5 tonnes, any prospect of ignition off site is to be ignored. However, all these figures are to be regarded as judgement factors for which there is no objective evidence.

Davenport's list of seventy-one vapour cloud explosions[5] contains five possible cases of off-site ignition and explosion. This suggests that such events might happen in some 7 per cent of cases. Such assumptions, combined with the data given in Table 7.2, lead to the off-site conditions shown in Table 7.3.

If material drifts off site unignited, the chances of it going in a particular direction and reaching a specific destination will depend on the wind and weather conditions as described in the previous chapter on dispersion. If the material ignites on site but does not explode, the calculation usually assumes that the drifting cloud will have the same downwind range as if it were unignited. The most likely scenario is for on-site explosions followed by fire, and the

Table 7.2
On-site chances of ignition and explosion

For release sizes of or less than (tonnes)	Chances of ignition	Chances of ignition and explosion
5 000	1.00	0.70
2 000	0.94	0.50
1 000	0.80	0.40
500	0.60	0.32
200	0.45	0.20
100	0.38	0.15
50	0.27	0.11
20	0.17	0.06
10	0.11	0.04
5	0.06	0.025
2	0.04	0.012
1	0.022	0.006
0.5	0.013	0.003
0.2	0.006	0.001
0.1	0.001	0

Table 7.3
Off-site chances of ignition and explosion

For release sizes of or less than (tonnes)	Chances of edge:		Chances of central:		
	ignition	ignition and explosion	ignition	ignition and explosion	Chances of no ignition
5 000	0	0	0	0	0
2 000	0.05	0.04	0.01	0.01	0
1 000	0.14	0.13	0.04	0.03	0.02
500	0.07	0.06	0.03	0.02	0.30
200	0.05	0.03	0.01	0.007	0.49
100	0.05	0.02	0.01	0.005	0.56
50	0.04	0.01	0.01	0	0.68
20	0.03	0.01	0.01	0	0.79
10	0.02	0.01	0.01	0	0.86
5	0.01	0	0	0	0.92

calculations for consequential fatalities are explained in the next chapter.

For off-site ignitions and explosions the main causes of death will be radiation and fire since the chance of an unconfined vapour cloud explosion generating a sufficiently high overpressure to cause primary death is remote. A typical estimate of the off-site fatalities involves the following steps:

1. calculating the maximum downwind range in each of eight 45-degree sectors, starting with the north (i.e. sector 1 = N);
2. assuming that no more than 20 per cent of the released quantity will be within flammable limits and that the maximum achieved cloud size is only 50 tonnes due to dispersion effects;
3. assuming that the boundary limit for fatalities due to fire alone will correspond to a radiation level of 12.6 kW per square metre;
4. calculating fatalities due to blast from an explosion alone following the damage relationship explained in the next chapter;
5. for ignition and explosion, the consequential fatalities per release are based on the firestorm data given in the next chapter, namely 36 per cent of the exposed population (obtained by multiplying the population density by the area derived from application of the previous boundary limit);
6. for fire only, the fatalities are more simply assumed to be one-fifth of the casualties caused by fire and explosion;
7. the number of fatalities per annum for a given mass release is then obtained by multiplying the derived frequency of release by the ignition/explosion chance, by the chance of the weather being suitable for the downwind range to overlap the population edge, and by the number of consequential fatalities per release.

Illustrations of the relative contributions considered so far

Before proceeding further down the consequence chain, where the difficult subject matter of possible damage to people is examined, two simple examples of the above calculations will be given. These examples have been chosen to illustrate the relative importance to the risk estimate of the factors introduced in this and the two preceding chapters. Both examples are confined to assessments which present the effects in terms of fatalities of an inflammable hydrocarbon release.

The first example postulates a release of hydrocarbon which may lead to fire and explosion on site. Separated from the site by a 'safety' distance of 50 metres there is an off-site population with a density of 600 people per square kilometre. This is an average figure for large residential areas with appreciable open spaces; population in specifically built-up areas can easily exceed 10 000 people per square kilometre. The problems of estimating population density are examined in Chapter 9.

In Table 7.4 six sets of data are given for each of three differing 'slopes' of the release equation explained in Chapter 5, and two differing 'safety' distances. The data provides the number of fatalities for various sizes of release taking into account the release frequency, the effect of the 'safety' distance and the chances of ignition and explosion. Figure 7.1 provides a graphical illustration of the data.

The maximum number of fatalities per event is not large, but obviously is linearly related to the population density. Increasing the 'safety' distance by 50 per cent reduces this number by a factor of three. It will be seen that even a 50-metre 'safety' distance eliminates the effect of the smaller releases, which are in any case discounted by their reduced likelihood of ignition and explosion.

The decrease in frequency of release with increase of size, coupled with the effective maximum of 50 tonnes, markedly reduces the effect of the larger releases. In consequence marginal reductions in the maximum size of the release do not have much effect.

The summated deaths per annum (Figure 7.1) are very small, at worst around four in a hundred years, but the graphs clearly show the effects of changing the 'slope' of the release and the safety distance. Changes in the population density or explosion chances are likely to be less dramatic.

It has been shown that for hydrocarbon vapour releases, ignition precedes and is more likely than explosion. Thus, it often happens that there is fire without explosion so that death and destruction from fire is more likely than from explosion alone, although explosion frequently follows fire. On-site fires are unlikely to cause many deaths off site since the intensity of thermal radiation falls off very rapidly with increasing distance from the source. More to be feared are burning vapour clouds, which may drift in the wind

Table 7.4
Off-site fatalities from on-site explosions

Event data			Safety distance 50 m Fatalities		Safety distance 75 m Fatalities		
Charge release tonnes	Frequency × 10^9	Explosion chance × 1000	per event	per year × 10^9	per event	per year × 10^9	Slope
5000	80	700	18	993	7	385	
2000	161	500	18	1443	7	561	
1000	274	400	18	1955	7	755	
500	468	320	18	2669	7	1035	
200	947	200	14	2621	4	849	
100	1613	150	5	1145	0	0	
50	2750	110	1	332	0	0	
20	5564	60	0	0	0	0	−1.3
10	9483	40	0	0	0	0	
5	16162	25	0	0	0	0	
2	32703	10	0	0	0	0	
1	55738	6	0	0	0	0	
1	94998	3	0	0	0	0	
0	192231	1	0	0	0	0	
0	327631	0	0	0	0	0	
5000	80	700	18	994	7	386	
2000	199	500	18	1783	7	694	
1000	398	400	18	2835	7	1095	
500	796	320	18	4542	7	1763	
200	1991	200	14	5511	4	1785	
100	3981	150	5	2827	0	0	
50	7962	110	1	962	0	0	
20	19905	60	0	0	0	0	−1.0
10	39811	40	0	0	0	0	
5	79621	25	0	0	0	0	
2	199054	10	0	0	0	0	
1	398107	6	0	0	0	0	
1	796214	3	0	0	0	0	
0	1990536	1	0	0	0	0	
0	3981072	0	0	0	0	0	
5000	80	700	18	995	7	386	
2000	295	500	18	2644	7	1029	
1000	794	400	18	5657	7	2185	
500	2138	320	18	12198	7	4733	
200	7916	200	14	21918	4	7100	
100	21309	150	5	15130	0	0	*cont'd.*

Table 7.4 *concluded*

Event data			Safety distance 50 m Fatalities		Safety distance 75 m Fatalities		
Charge release tonnes	Frequency × 10^9	Explosion chance × 1000	per event	per year × 10^9	per event	per year × 10^9	Slope
50	57361	110	1	6932	0	0	
20	212373	60	0	0	0	0	–0.7
10	571667	40	0	0	0	0	
5	1538813	25	0	0	0	0	
2	5697331	10	0	0	0	0	
1	15336077	6	0	0	0	0	
1	41281659	3	0	0	0	0	
0	152842054	1	0	0	0	0	
0	411420301	0	0	0	0	0	

towards and over a residential area, and this provides the subject matter for the second example.

Figure 7.2 provides a graphic illustration of the average annual fatalities likely to be caused by a burning, drifting cloud that has been ignited on site. It is based on the estimate obtained from a simple computer model constructed in accordance with the steps previously enumerated. The model comprises eight 45° sectors, and the weather pattern is simplified into five combinations of Pasquill categories and windspeeds in metres per second: F2, E2, D2, D5 and C5. The frequency of these weather combinations in each of the eight sectors is based on information appropriate to Canvey, but in Figure 7.2 the results for one sector only are shown. The downwind ranges are based on the scaling law information given for propane in the previous chapter. Results are tabulated for three 'safety' distances – 50, 250 and 500 metres – but graphic illustrations are shown for only two – 250 and 500 metres.

The shape of the fatality rate/log tonnes curves is broadly similar to those of Figure 7.1, and is influenced by the same considerations of release pattern, 50-tonne maximum and safety distance, whilst the ignition frequency, although greater than that for explosion, is similar in order of magnitude and in its relationship to the mass released. The potential spread of the damage is significantly wider

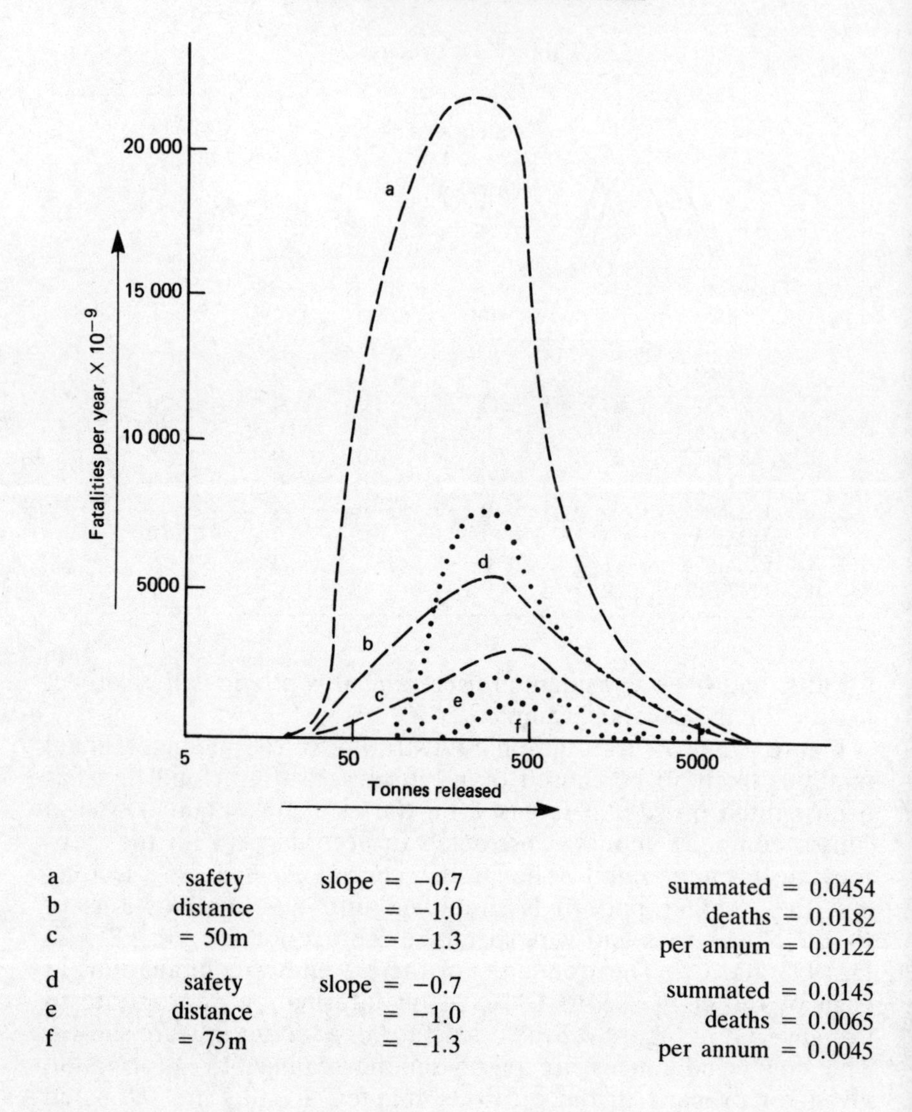

Figure 7.1 *Off-site fatalities from on-site explosions*

than for the blast damage cases of Figure 7.1, but very dependent on the weather in matters of detail. Up to 250 metres, a contribution is made from each of the five weather combinations, and the sector most at risk is in line with the direction of the prevailing wind. Beyond this distance the larger releases under light winds present

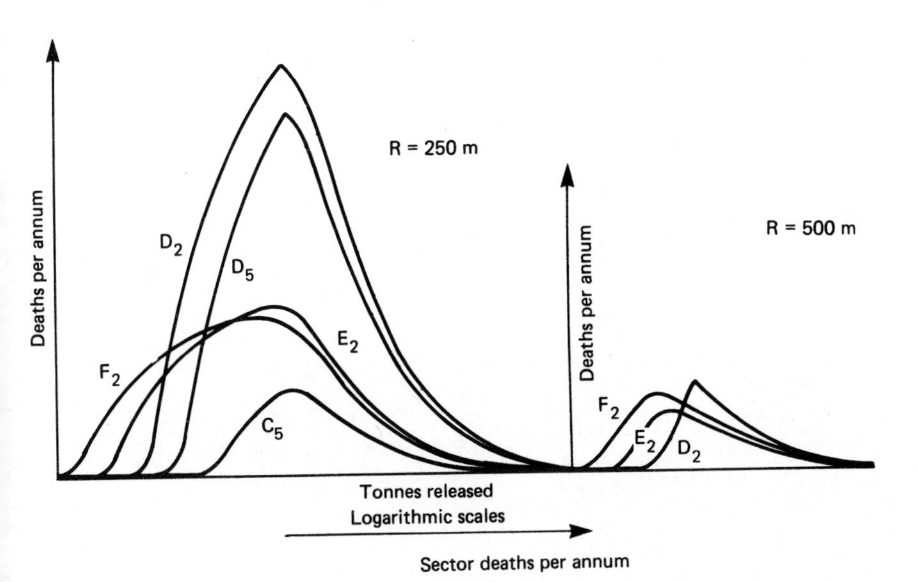

	Sector deaths per annum		
Pasquill categories	beyond 50 m	beyond 250 m	beyond 500 m
F_2	.0010	.0007	.0004
E_2	.0011	.0007	.0003
D_2	.0035	.0021	.0004
D_5	.0037	.0019	.0000
C_5	.0010	.0003	.0000
Summated sector deaths/y	.0103	.0057	.0011

Figure 7.2 *Off-site deaths per annum*

the main hazards, the frequency is much reduced and more randomly oriented. At 500 metres the contributions from D5 and C5 have disappeared.

The curves are merely indicative of average consequences, and a real situation would be further complicated by the significant

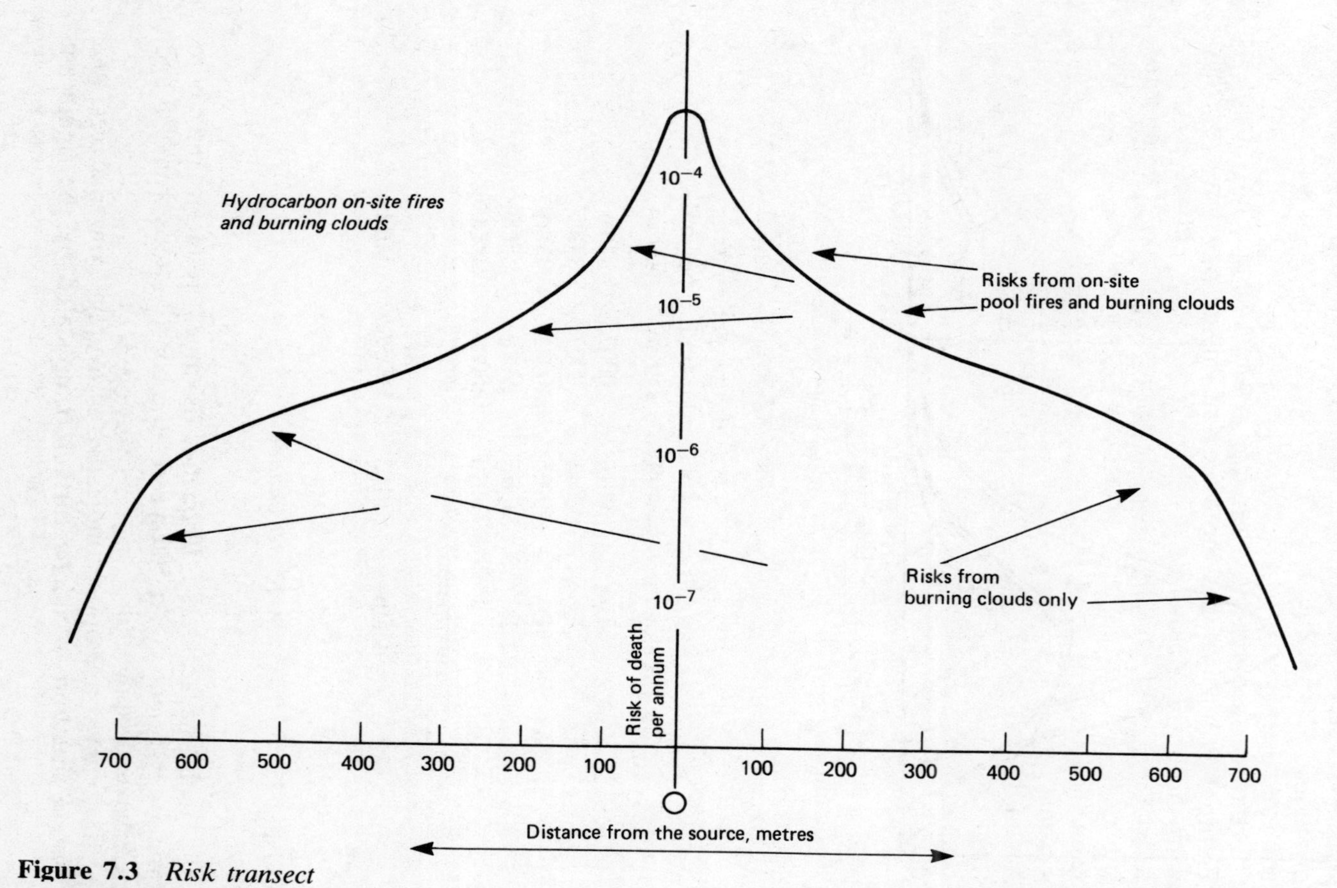

Figure 7.3 *Risk transect*

variations in the distributions of the target populations around the source. As has been remarked in Chapter 1 there are conceptual problems in the summation of disparate components of societal risk, and perhaps the portrayal of individual risk shown in Figure 7.3 is more meaningful. It is a risk transect similar in concept to Figure 1.1, but includes the effects of on-site pool fires together with the burning cloud data of Figure 7.2, but without the need for data on population. It gives emphasis to the much greater on-site risks, and to the rapid improvement with increasing distance from the source.

Some remarks about the relative importance of fire and explosion have already been made in Chapter 2 on the lessons to be learnt from case histories. Further comment will be made in the next chapter which examines the various damage relationships in some detail.

References

1 Wiekema, B. J., *Analysis of vapour cloud accidents*, Proceedings of the Fourth Euredata Conference, Venice, 1983.

2 Badoux, R. A., *Some experiences of a consulting statistician in industrial safety*, Proceedings of the Fourth Nat. Ref. Conference 3B, 1983, 5. 1.

3 Kletz, T. A., Second International Symposium on Loss Prevention, p. 50, *I. Chem. E. Symposium Series*, 1968.

4 *Canvey Second Report*, p. 37, London: HMSO, 1981.

5 Davenport, J. A., *A study of vapour cloud incidents – an update*, I. Chem. E. Fourth International Symposium on Loss Prevention, Harrogate, September 1983, c1.

8 The damage relationships

It was stated in the beginning of Chapter 1 that the most readily understood form of risk is the chance of death to an individual. This simple definition of risk had to be immediately qualified in respect of the possible location of the individual in time and space, further distinctions were then made between instant and delayed death and mention was made of the need to take account of injuries and fates 'worse than death'.

The point was then made that the calculation of individual risk involves working through the long chain of events leading to the release and dispersion of hazardous material to obtain the chance of death occurring to an individual permanently located at the defined point in space away from the source of the release. The final element in the long chain is the damage relationship, which relates the chances of damage (death or injury in the case of individuals) to some dependent variable of the release, such as concentration in the case of a toxic release or overpressure in the case of an explosion.

Whilst death is the extreme form of damage to an individual, many other forms of damage may have to be considered when assessing the possible consequences of industrial hazards. Thus, for example, in the case of possible hazards from the disposal of highly toxic or radioactive material, consideration may have to be given to the pollution of foodstuffs or drinking water. In the case of the storage or transport of chemical hydrocarbons, consideration may have to be given to possible fire damage or property. A full consideration of damage relationships must therefore encompass a very wide spectrum of causes and effects.

In so far as the damage to individuals is concerned, some categories of instant death can be related to other forms of death or to injury or illness, but this is not possible in many cases due to the lack of adequate data. Even the establishment of a precise cause and effect relationship for instant death can be considerably difficult. This difficulty is exemplified by the inherent variation in response between one individual exposed to a hazard and another. It is a characteristic of people that some have a capacity to withstand or recover from damage which may be lethal to others. This variation is a feature of the records of case histories in peacetime as well as in war. Over many years it has been repeatedly observed in animal tests upon lethal effects, whether they be about inhalation toxicity, overpressure or nuclear radiation.

It is convenient to categorize the damage relationships under the headings of:

1 explosion;
2 fire;
3 nuclear radioactivity;
4 toxicity.

This simple list is soon complicated by the fact that in each case, even for instant death, there are numerous routes by which the final consequence can be related to the prime cause. This is readily illustrated in the examination of the first item, death from explosion or blast.

Damage from explosions

An explosion involves a process whereby a pressure or blast wave is generated in air by a rapid release of energy. The front of this wave can cause damage to the objects it impacts as it passes through the air.

Explosions may cause death and injury in a variety of ways but it is a common practice to relate the blast overpressure and duration time to the quantity of explosive material first and, second, to try to relate the pressure-time history of the blast wave at a specified distance from the explosion source to a damage criterion such as percentage fatalities.

Much of our understanding about the damage caused by explosions comes from experience of wartime bombing and from peacetime field experiments with condensed phase explosives. It is therefore convenient to describe the effects in terms of a TNT explosion and to relate the effects of other explosive substances to TNT by means of a conversion formula. This will be explained later.

The characteristics of a blast wave produced from a TNT explosion are relatively well understood and a full account is given in a book by Baker *et al.*[1] The magnitudes of the key parameters such as overpressure, duration time and impulse at known distances from the source can be related to the mass of explosive by means of a scaling law. Perhaps the earliest scaling law is Hopkinson's,[2] restated by Zuckerman in 1940: 'if a given positive pressure is caused by a given amount of explosive at a given distance, the same pressure will be experienced at twice that distance only when the amount of explosive is increased eight times'.[3]

When TNT explodes, the solids are converted into gases at a pressure up to perhaps 650 psi (4550 kPa). As a result a blast wave is generated in the surrounding air which has both a positive and a negative component. For a 70 lb (32 kg) charge (typical of UK bombs used at the outset of World War 2), Zuckerman showed that the positive pressure pulse lasted about 3 milliseconds and the suction pulse about 30 milliseconds. More recent work has established that the overpressure duration for a 1-tonne TNT explosion at 5 psi (35 kPa), is 40 milliseconds, and for a 2500-tonne explosion, 500 milliseconds. For a million-tonne nuclear burst the corresponding duration would be 2.5 seconds.

It is important to take time into account as well as overpressure when estimating the damage that can be caused by a blast wave, and as a first approximation it is convenient to formulate the possible damage in terms of impulse, simply expressed as half the product of pressure and time.

The experimental work has demonstrated a significant scatter, due to differing chemical explosives, differing containers, differing geometric aspects, and so on, and a plot showing the differing results is given in Baker's book.[4] However, by taking a mean line through the Baker curves and making an allowance for the reflection of the blast by the ground a reasonably representative set of curves can be obtained and this is illustrated in Figure 8.1. It shows the overpressure from a condensed phase explosion as a function of explosive mass and distance.

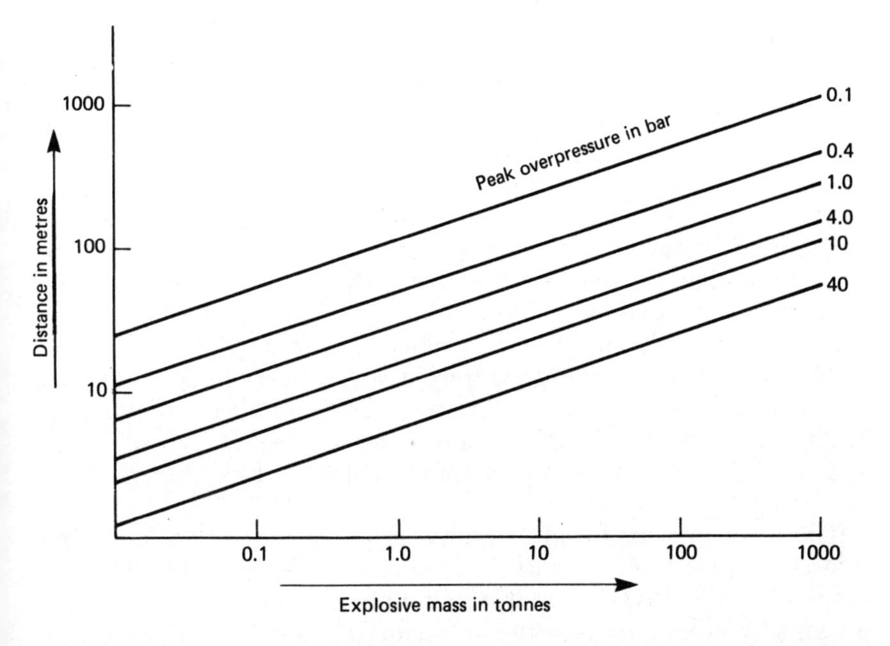

Figure 8.1 *Overpressure as a function of explosive mass and distance*

The classification of blast casualties

Zuckerman's work was sponsored by the UK Ministry of Home Security in 1939–41, and was directed towards establishing criteria for the construction of air raid shelters. This was needed because at that time more tons of high explosive were dropped on a civilian population each night than throughout the four years of World War 1.

To gain data directly relevant to human beings, Zuckerman spent much time examining air raid casualties, covering both survivors and autopsies. Some remarkable escapes drew attention to the great variability between individual responses, but nevertheless Zuckerman and his co-workers set about a series of experiments involving tethered animals (goats, rabbits, rats and so on). These were exposed to blast from bombs (most of the work was done with 70 lb bombs in paper containers to avoid the effect of splinters, but some work was done with 500 lb RAF GP bombs, and some splinter experiments were undertaken by firing high velocity steel balls into rabbits' legs). The work sought to establish the prime parameters of the blast mechanism as a direct cause of death, to draw up a specification for air raid shelters.

The autopsies had shown that some deaths were associated with severe lung haemorrhage, and there was controversy as to whether this was caused by the overpressure wave entering the breathing tract or whether it was due to external pressure on the body. Accordingly, some of the experiments involved protecting the animals' bodies in containers so that only the head and breathing passages were exposed. This work established that the lung haemorrhage was caused by external pressure on the body.

Although it was clear that different species vary in susceptibility and that there was significant variation between individual members of the same species, a relationship between average body weight and the percentage fatality was found which enabled the results to be extrapolated to man.

The work provided useful results for the shelter programme but it also gave fresh impetus to the users and designers of bombs. By detonating the British 500 lb GP bombs amongst live goats in a pit and by cross-checks with the results of raids on British cities, Zuckerman and his co-workers were able to establish that the lethal pressure to an average man for this type of bomb was between 400 and 500 psi. Further calculations by operational research departments showed that the RAF's projected area offensive against Germany would be a futile affair with such a size of bomb, and there came to be developed 4000, 8000, and finally 12 000 lb 'Blockbuster' bombs for the area offensive.[5]

Zuckerman also evolved a classification of death and injury due to blast action.

1 Being directly affected by impact of the blast wave, being directly hit by the bomb or by splinters from the bomb case, being burnt by flames from the bomb, or being poisoned by the carbon monoxide liberated from the bomb in a confined space.
2 Being bowled over against a hard surface by the blast wave or by a splinter.
3 Being hit by secondary missiles, having walls or ceilings fall upon one, falling due to a collapsed floor.
4 Effects due to asphyxiation and burning in a trapped position because of secondary fires.

Zuckerman's work on autopsies showed that it was difficult to distinguish the effect of an atmospheric blast wave acting upon the body and the effect of impact on the body by missiles or by translation of the body on to a hard object. The impression gained from reading Zuckerman's papers and the discussion of them is that in air raids death or injury due to lung damage was extremely rare.

Primary blast relationship

For the purposes of hazard assessment it is proposed to classify casualties under just two headings:

1. *primary* – being directly affected by impact of the blast wave or from fragments of the explosive's casing;
2. *secondary* – being hit by secondary missiles, walls, ceilings, and so on, or falling due to a collapsed floor or being buried.

A considerable amount of work has been carried out involving animal tests which has provided a good understanding of the damage relationship for the primary category. It may be divided into two parts: those which are associated with a stationary body being hit by the blast wave, and those which result from a body being bowled over in the wind of a blast wave or hit by a primary splinter.

Probably the most comprehensive experimental investigation into the effect of a blast wave upon a stationary body was that carried out by the Lovelace Foundation at Albuquerque, New Mexico, for the US Atomic Support Agency. It has been described in many reports and papers,[6] but those of Bowen, Fletcher and Richmond are particularly significant. Tests were carried out on 2097 animals in thirteen different species ranging from mice to steer, with blast wave pressures up to 1680 psi (11 600 kPa or 116 bar), and duration times from 0.3 to 400 milliseconds. Table 8.1 gives typical averaged results for rabbits, goats and sheep.

From such data it is possible to derive a set of curves showing the relationship between impulse and duration time for 50 per cent lethality of the different species at standardized bodyweights. Figure 8.2 shows curves for the three species in Table 8.1 and a fourth curve for a standardized 70 kg man, constructed by extrapolation; the degree of extrapolation required is slight.

From the data in Figures 8.1 and 8.2, and the scaled relationship between duration time and explosive mass, it is possible to derive a relationship between mass of explosive and distance for 50 per cent lethality to a 70 kg man and this is illustrated in Figure 8.3.

Table 8.1
Lovelace field experiments

Species		Average weight (kg)	Average pressure (psi)	Average time (milliseconds)	Mortality killed/tested
Sheep	1	52.5	164	3.18	31/65
	2	54.0	52	211	14/34
	3	52.6	1 307	0.30	7/22
Goats	1	22.7	111	3.81	7/15
	2	23.2	60.9	17.0	8/15
	3	21.7	55.8	38.7	16/28
	4	20.5	51.5	400	13/30
	5	24.7	295	1.2	6/12
Rabbits	1	1.9	78.9	1.12	32/49
	2	1.9	44	2.79	25/50
	3	3.7	25.7	354	17/32

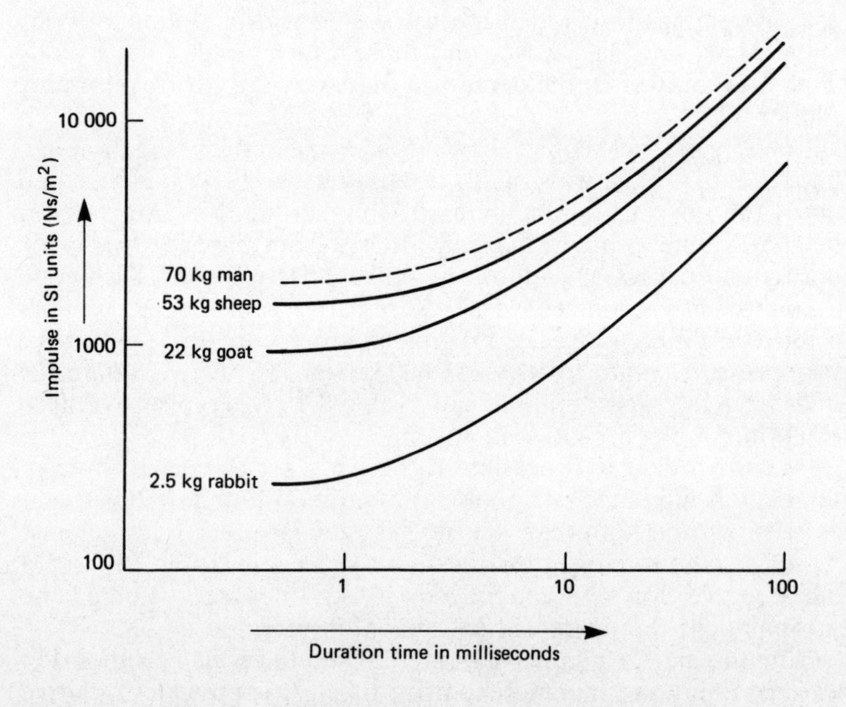

Figure 8.2 *Impulse and duration time for 50 per cent lethality*

Figure 8.3 also illustrates a similar relationship for primary deaths caused by whole body translation. Glasstone and Dolan describe experimental work on this mode using animals and dummies and propose a scaling law:[7]

$$d = d_r \times W^{0.4}$$

where d = distance for 50 per cent casualties;
W = mass of explosive in kilotonnes;
d_r = distance for a 1-kilotonne explosion.

The two curves are almost identical, and can be represented by the following formula. It can be regarded as the 50 per cent lethality contour for primary blast deaths, giving the distance in terms of the primary independent variable, explosive mass.

$$L = \frac{2 \times 10^{-2} \times T^{1/3}}{[1 + (20/T)^2]^{1/6}}$$

where T is explosive mass (tonnes) and L is distance (km).

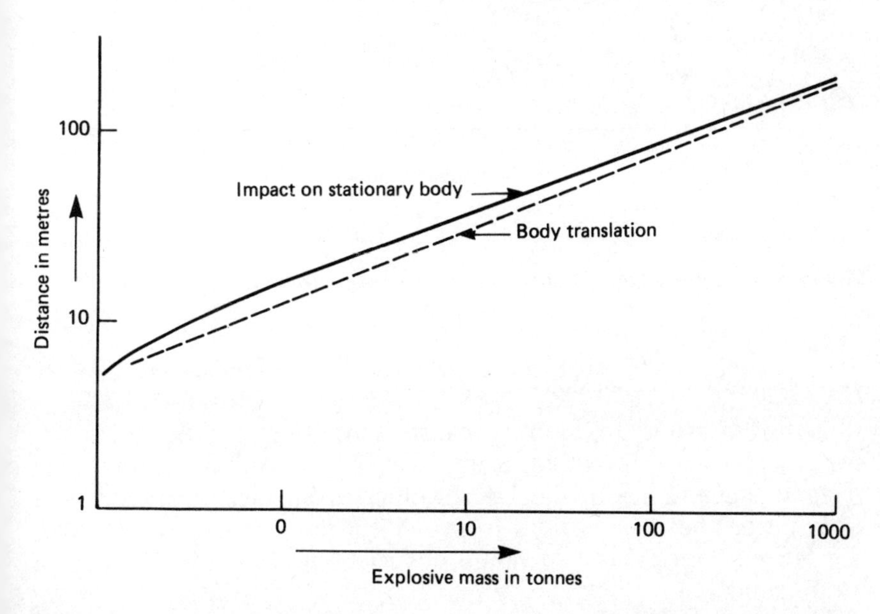

Figure 8.3 *Point source explosions primary blast deaths 50 per cent fatality to 70 kg man*

It is also possible to derive fatality relationships in probit form from the animal data. Figure 8.4 gives an indication of the prediction for a 70 kg man in terms of overpressure and percentage mortality. Short-duration time pressure waves (associated with small explosions) require a much higher pressure for lethal effect than do the longer duration times (associated with larger explosions).

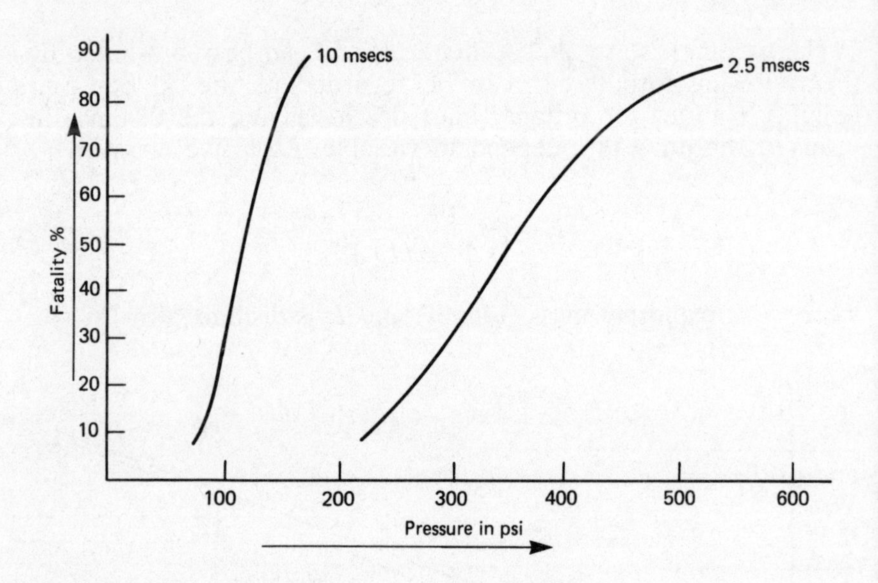

Figure 8.4 *Percentage fatality versus overpressure*

Secondary blast relationships

It is not possible to identify cause and effect relationships for secondary categories in the same way as for primary categories. Instead the findings of the UK Explosives Storage and Transport Committee[8] have been used to provide a risk contour which gives the 50 per cent chance of home destruction from various sizes of TNT explosion.

The relationship between home destruction and explosive mass is subject to considerable variation because:

1 there are differences in the blast wave overpressure from one explosion to another, as has already been noted;
2 the blast wave from a point source cannot radiate symmetrically due to variations in terrain and in obstructions;
3 there are inherent differences in the construction and aspect of buildings.

However, it is possible to define a category of damage to residential property in the UK, which on average represents it being made uninhabitable as a result of the blast from an explosion. Such a category can be used to identify a contour where, as a result of a specified explosion, the number of uninhabitable houses outside the contour are balanced by the number of inhabitable houses inside the contour. The contour can be calculated from the following formula:

$$L = \frac{13.2 \times 10^{-2} \times T^{1/3}}{[1 + (3.175/T)^2]^{1/6}}$$

where T is explosive mass (tonnes) and L is distance (km).

As a first approximation this contour can be considered as equivalent to a 50 per cent damage contour, similar in concept to the 50 per cent primary damage lines of Figure 8.3. It has been derived from an analysis of 110 reports of accidental explosions, and of the damage to houses in the UK from air raid bombings in World War 2. The mass of explosive ranged from 300 lb to 5.3 million lb, whilst the explosive type included dynamite, TNT, Torpex and guncotton. The relationship between the primary and secondary contours is shown in Figure 8.5.

To form an estimate of the number of people likely to be affected by secondary blast action, an estimate is required of the ratio of deaths and injuries to the homes made uninhabitable. Since the uninhabitable homes outside the contour are balanced by the inhabitable homes inside, it is only necessary to know the number of homes inside the contour.

An estimate of the relationship between uninhabitable homes and casualties has been formed from a review[9] of the following categories of event:

1 natural disasters, including:
(a) tornadoes;
(b) earthquakes;

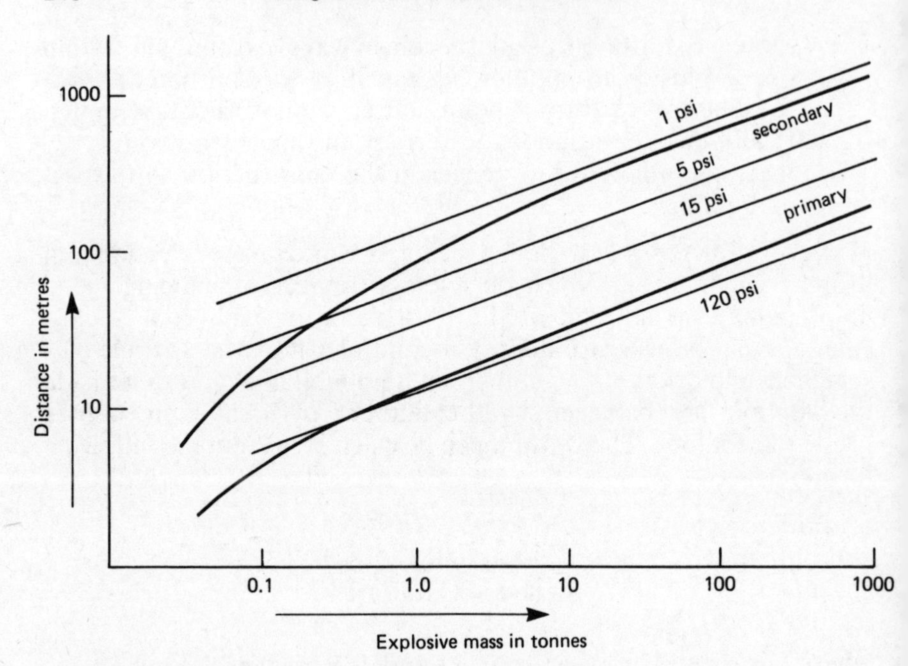

Figure 8.5 *Primary and secondary reference lines (blast damage relationships)*

2 man-made disasters, including:
  (a) chemical explosions;
  (b) air raids over Germany and the UK;
  (c) the V2 attacks;
  (d) domestic gas explosions in the UK.

From the review of the man-made disasters, the ratio of deaths to uninhabitable houses is of the order of 1:10–15. In the case of natural disasters the death toll is rather less than this, presumably because they do not often occur without warning. For the purpose of hazard assessment a ratio of one death to every ten occupied homes made uninhabitable is proposed.

The ratio of deaths to injuries is of the order of 1:10 for industrial chemical and gas explosions; the wartime bombings resulted in rather more deaths in relation to injuries.

An average figure for home occupancy in the UK is 2.5, so that the number of casualties in a 45° sector is then given by the formulae:

$v = \pi R^2$ where R is the distance from the hazard source to the population boundary (v is the clear space)

$q = (x - v)/8$

$s = (z - v)/8$

where x and z are the primary and secondary damage zones around the hazard source

$n = D \times (q + s/25)$ where D is the population density and n the number of fatalities.

Because of the much larger 50 per cent damage radius there will be more fatalities arising from secondary effects than from primary even though only one fatality is to be expected for every ten homes destroyed. This is illustrated in Table 8.2, which also shows the very small correction that would be made to allow for double counting of the two fatality totals are simply added together.

From studying the findings of previous workers, the conclusion has already been drawn that far fewer deaths are caused from blast relative to deaths caused by fire. Deaths from primary blast alone are very rare. This, for example, was the conclusion of Zuckerman's work during World War 2 on air raid casualties, whilst the more recent TNO report[10] on the San Juan Ixhuatepec LPG incident in Mexico City on 19 November 1984 concludes that there was only minor damage from the blast wave, but that the greatest damage was probably caused by fire and from explosions of the gas that accumulated inside houses. The distance from the storage area to the far border of the area where most of the houses were destroyed was about 300 metres. The nearest houses were about 100 metres

Table 8.2
Explosions at random in populated area (4000 persons per square kilometre)

	Primary deaths		Secondary deaths		
Explosive mass (tonnes)	Radius (metres)	Number	Radius (metres)	Number	Total deaths (corrected)
1000	170	363	1310	862	1210
100	80	80	607	186	263
10	34	14	256	33	47
1	13	2	88	4	6
0.1	5	1	19	0	1

from the edge of the storage area. Within this area approximately 500 people were killed and altogether 7000 were injured. The majority of those killed were still asleep when they were surprised by fire, and did not survive because of direct flame contact, heat, fumes and the lack of oxygen. The radius of 300 metres agrees reasonably well with the TNO-calculated fireball radius that may have occurred, and the percentage mortality of the people living within this seems to have been as high as that for the German cities subjected to firestorms in the RAF air raids in World War 2.

The TNT-equivalent concept

The data given above relates to point source TNT explosions, since it is convenient to conduct damage experiments on this basis and there is also much data available relating wartime damage in this way. The usual approach to estimating blast effects from industrial chemical explosions is to form a TNT equivalent.

To obtain the TNT equivalent the estimated weight of the released material has to be multiplied by two factors: the first provides a ratio between the energy per unit weight of TNT and the released material; the second is an arbitrary efficiency factor. Table 8.3 gives suggested values for these factors, reproduced from a recent report of the MHAP working party,[11] for a few common industrial chemicals.

For hydrocarbons an energy ratio of 10 applies, and the MHAP working party recommends an efficiency factor of 0.042. Thus, for example, a 100-tonne tank of liquid propane might on failure produce by flash and spray action a vapour cloud containing 64 tonnes of propane, equivalent to 27 tonnes of TNT. Thus, the explosive mass of hydrocarbon is equivalent to approximately 50

Table 8.3
Some suggested values for TNT equivalence

Material	TNT equivalence	Efficiency factor
Ethylene	0.6	0.06
Other hydrocarbons	0.4	0.04
Ethylene oxide	0.6	0.10
Propylene oxide	0.4	0.06
Vinyl chloride	0.2	0.04
Methyl chloride	0.1	0.04

per cent of the actual release. This provides a basis for a conservative estimate of the consequences of an on-site explosion where there may be an element of constraint or confinement on the escaping material. Where, however, an unconfined vapour cloud drifts off site, much lower TNT equivalents are thought to apply.

Davenport[12] estimates that the TNT equivalence from his portfolio of explosive vapour cloud case histories is an average 2 per cent, but there is a wide variation from 50 per cent to less than one per cent.

It has been estimated that only 15 per cent of the volume of a drifting cloud is likely to lie within explosive limits, and it is only this portion of the cloud which is likely to set up an overpressure and cause blast damage. There is grave doubt whether the TNT-equivalent concept is valid for a drifting cloud, and none of the experimental or case history evidence so far obtained has suggested the creation of an overpressure greater than 3 psi or 0.3 bar.

Very few leaks are instantaneous and for large quantities it is considered that by the time the fiftieth tonne has emerged, the first tonne will have dispersed below explosive limits. It is therefore suggested that a limit of 50 tonnes is placed on the TNT equivalence of the explosive material, which is otherwise 50 per cent of the mass released for on-site explosions, but only 10 per cent for clouds which drift off site prior to explosion.

To summarize, the special features of the concepts proposed for the assessment of explosion hazards include:

1 The concept of a primary damage contour within which those who survive are balanced by those beyond who die.
2 The concept of a secondary damage contour within which those residential homes which escape destruction are balanced by those beyond which are destroyed. A relationship of one fatality for every ten homes destroyed is assumed together with an average night-time occupancy of 2.5 per home.
3 Explosions which occur on site are distinguished from those which take place as a result of a vapour cloud exploding off site. On-site ignition is distinguished from off-site ignition.

Damage from fire and radiant heat

Damage from fire and radiant heat can occur in three main ways:

1 ignition at the factory or terminal causing a large but non-explosive fire on site;

2 ignition of a rapidly expanding cloud of vapour leading to a 'boiling liquid expanding vapour explosion' (BLEVE) and/or fireball;
3 ignition of a drifting cloud of dense gas off site to form either a flash fire or an explosive fireball.

For people caught in the open there is a risk of death or injury from thermal radiation. There is evidence that this effect is time-dependent, and the probit equation is likely to be of the form:[13]

$$\mathrm{Pr} = \mathrm{a} + \mathrm{b} \log I^{n} t$$

where Pr = probit;
a, b = constants;
I = radiation intensity in kilowatts per square metre;
n = 4/3;
t = time in seconds.

The limiting value of I for periods longer than 45 seconds may be estimated at 4 for human beings, and for a 1000-tonne pool fire this suggests a limiting range of the order of 800 metres, but there is some uncertainty about the source heat flux from such fires and the obscuring effect of smoke which may significantly reduce the hazard range. Only a small proportion of people are likely to be out of doors under normal conditions, and the limiting value of I for 'secondary' fires, that is radiation-ignited buildings, is 13 kilowatts per square metre, which gives a limiting range for people inside such buildings of the order of 500 metres, but this is subject to the same uncertainties in the estimated heat flux.

A common rough estimate for the hazard range for large pool fires on site is about two pool diameters, and for large drifting vapour clouds 1.5 cloud diameters. A rapid release of flammable gas can cause a fireball in the air and in some cases this will be associated with a source heat flux very much higher than will be the case for pool fires, giving the possibility of large hazard ranges. On the other hand, it has been suggested that there is likely to be a limiting value of 50 tonnes for the mass in such a cloud and only 20 per cent of this may be within inflammable limits. An approximate expression[14] for fireball hazard range is:

$$R = 0.11 M^{\mathrm{n}}$$

where R = range in kilometres;
M = mass of gas in fireball in tonnes;
$n = 0.4$.

This gives a limiting range of 500 km for the largest releases. It will be noted that this formula is very similar to the one used in the dispersion scaling law.

People may escape from a burning building, but they may be trapped and overcome by smoke and fumes. They may also take refuge in cellars but the chances of death are increased if the fire is accompanied by explosions which severely damage the housing structures. Such an explosion may be central to the vapour cloud and initiated on site or, as at Mexico, be severally dispersed among the various buildings.

Because of the varied nature of the causes of death from such fires it is difficult to formulate a damage relationship from a single primary independent variable such as mass of the gas released, and not much use can be made of the relationship involving radiation alone. As a first approximation, a simplified approach may be followed involving these three basic tenets:

1 for on-site ignitions followed by fire without explosion, the hazard range is made 100 per cent greater than the radius of the vapour cloud whose volume is given in Table 6.9 (this would be 260 metres for 1000 tonnes of LNG);
2 for on-site ignition and explosion of rapidly expanding clouds leading to a BLEVE and fireball, the TNO model[15] is used, namely $D = 6.48\ M^{0.325}$;
3 for off-site ignitions the maximum mass of vapour in the cloud is put at 50 tonnes and the hazard range given by the dispersion scaling law given in Chapter 6.

In the first case the chance of ignition only is taken from Table 7.2. In the second case the chance of ignition followed by explosion is also taken from Table 7.2. In the third case the chance of off-site ignition is taken from Table 7.3. This chance has also to be multiplied by an appropriate weather chance.

Table 8.4 compares the hazard distance obtained in the three ways given above for equal opportunities of released material. An estimate of the individual risk from fire and radiation can be made by adding the risk from each of the three cases together. If this is done the combined risk is about an order of magnitude greater than that from blast for the same release pattern.

Table 8.4
Comparative hazard ranges

Tonnes released	Hazard range (km) Case 1	Case 2	Case 3
1000	.260	.577	.525
500	.206	.461	.525
200	.152	.342	.525
100	.120	.273	.525
50	.096	.218	.525
20	.050	.162	.364
10	.038	.129	.276

Nuclear radiation damage

Although the number of early deaths caused by accidents in the nuclear industry has been quite negligible in comparison with other accidents, the public fears the nuclear industry because of possibly horrendous delayed deaths and unimaginable genetic effects. These are largely fears of the unknown, since although there have been extensive follow-up investigations of the offspring of survivors from the Hiroshima and Nagasaki explosions, there has so far been a failure to establish any statistically significant result in so far as genetic effects are concerned.

Explanations of nuclear risk have not been made easier by a bewildering array of scientific terms used in connection with the physics of radioactivity. The basic concept, however, is to define an amount of radiation which will absorb a specified amount of energy in a given weight of human tissue. This is specified as the 'rad' (roentgen absorbed dose); a new unit, the 'gray', equals 100 rads. To measure the biological effect of the absorbed dose, the 'rem' (roentgen equivalent man) was in common use, but the new international standard is the 'sievert', which is equal to 100 rem. The rem is the amount of absorbed radiation dose corresponding to 100 ergs of energy in one gram of tissue. The gray is the absorbed dose of one joule per kilogram (a joule equals ten million ergs). Different types of tissue absorb different amounts of energy and α-rays are more penetrating than β- or γ-rays. The sievert is the absorbed dose in grays multiplied by the quality factor of the radiation involved. β- and γ-radiation have quality factors of one, whereas α-radiation has a quality factor of 20.

An instant dose of over 1000 rems will certainly by fatal. No early deaths are to be expected from instant doses below 100 rems, but some increase in the incidence of cancer has been detected at instant doses around 10 rems. At present there is insufficient data to establish any kind of threshold below which radioactivity can be safely absorbed by tissue, so that a basic assumption is made that the risk factor (health consequence per unit of dose) is constant irrespective of dose or dose rate. It will be understood, however, that the consequences predicted at low doses will be dwarfed by the consequences of other types of hazard. Table 8.5, based upon the total risk factor currently recommended by the International Commission on Radiological Protection (ICRP), provides some illustrations of the likelihood of delayed death from exposure to radioactivity, together with some comparative deaths from other causes.

It can be seen, for example, that the lifetime risk from lung cancer from an effective annual dose equivalent to 200 millirems is one in 5000, which is about 0.1 per cent of the total risk of fatal malignancy facing the average individual, or about 0.02 per cent of the risk of death from all causes. An annual dose of 200 millirems is about what the average inhabitant in the UK receives from all sources at the present time. A statistical average of the annual public exposure from the entire UK nuclear industry, mainly the disposal of nuclear waste, is 0.3 millirems.

The US National Academy of Sciences Advisory Committee on the Biological Effects of Ionizing Radiation[16] has concluded that,

Table 8.5
Lifetime risks of premature death

Cause	Lifetime risk
Overall risk of fatal malignancy	2×10^{-1} (1 in 5)
Lung cancer from 10 cigarettes a day	6×10^{-2} (1 in 17)
All accidents	2×10^{-2} (1 in 50)
Accidents in the home	9×10^{-3} (1 in 110)
ICRP risk of any cancer, 1 rem a year	8×10^{-3} (1 in 125)
Accidents on the road	8×10^{-3} (1 in 125)
ICRP risk of lung cancer, 1 rem a year	1×10^{-3} (1 in 1000)
ICRP risk of lung cancer, 0.2 rem a year	2×10^{-4} (1 in 5000)
All fire accidents	1×10^{-4} (1 in 9650)
All cancer from 0.3 millirems a year	24×10^{-7} (*c.* 2 in 1 100 000)

in humans, an average parental exposure of 1 rem (1000 millirems or 10 millisieverts) before conception is expected to produce 5–75 additional disorders per million live-born offspring in the first generation. The current natural incidence of human genetic disorders is approximately 107 000 cases per million live-born offspring.

The 200 millirems a year received on average by everyone in the UK is obtained from a wide variety of sources, as shown in Table 8.6. It is based on the National Radiological Protection Board's 1984 review and some of the data may be revised.

The smallest contribution itemized in the NRPB review comes from the nuclear industry and Table 8.6 suggests that even if the size of this industry were to be increased fivefold (it presently provides 20 per cent of the CEGB's supply) no great additional impact of radioactivity would result.

The largest single contribution comes simply from being indoors, where the average UK person spends 90 per cent of his time. There are, however, very large differences in the amount of radioactivity experienced in various parts of the UK. While for the country as a whole the average dose is around 80 millirems a year for radon decay products, the average annual dose in Cornwall is around 640 millirems, and in five per cent of Cornish dwellings the annual dose exceeds 2500 millirems. Nevertheless, in spite of the widely quoted risk parameters, based upon the hypothesis of linear response, there is yet no epidemiological evidence of actual harm to the inhabitants of these Cornish dwellings.

Table 8.6
Origin of average annual dose in the UK

Source	Annual dose (millirems)	% of total
Natural cosmic rays	30	14
γ-rays in the soil	40	19
Radon/thoron in buildings	80	37
From our bodies and food	37	17
Medication	25	11.5
Occupational, travel, etc.	2	1.0
Fall-out from weapon systems	1	0.5
Nuclear power, mainly waste	0.3	0.1

Major nuclear hazards

Until March 1986 the International Atomic Energy Agency was 'proud to be able to say that we had accumulated 3800 reactor years of experience without a single fatal radiation accident at a commercial nuclear power plant being reported and there had never been an accident with large-scale radioactive releases'.[17] The Chernobyl accident has drastically changed that situation. As a result some countries have modified their nuclear energy plans and others face demands that nuclear power be phased out altogether. But because the Chernobyl reactor is so different from the types favoured outside the USSR, it seems unlikely that anything learnt from Chernobyl can affect the risk assessment of nuclear power stations in the UK. It is, however, much to be hoped that the Soviet authorities will conduct comprehensive investigations into the health effects of the accident over the next thirty or forty years, preferably as an international study.

The greatest hazard for a nuclear reactor is a series of events which might impair the cooling of the core followed by a failure of the containment structure, leading to an escape of radioactive material which would be blown downwind over a densely populated region. The risk is calculated from equations which have the typical form:

$$\text{Risk} = f \times Ps \times Pc \times Pw \times C$$

where f = frequency of coolant failure;
Ps = probability of emergency supply failing;
Pc = probability of containment failure;
Pw = probability of wind carrying radioactive particles to a residential area;
C = consequence, for example number of deaths, based on the release magnitude, population density, and so on.

Figures 8.6 and 8.7 give the individual risk transect and societal risk estimates produced for the Sizewell 'B' PWR.[18] For purposes of comparison they have been superimposed upon information given in Figures 1.1, 7.3 and 1.5 respectively. For people living near to the station the individual risk is less than 2×10^{-9} per year, which adds no more than one part in 100 000 to the risk of everyday accidents in the UK, and is 5000 times better than the Royal Society's 'acceptable' risk figure. The ammonia and hydrocarbon

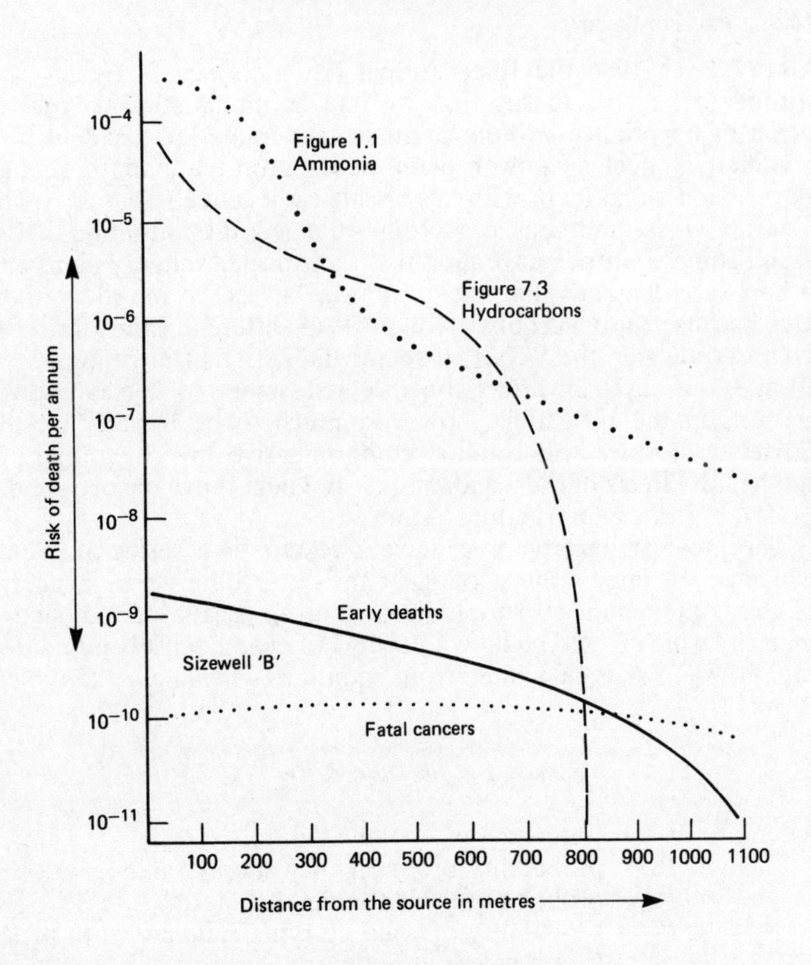

Figure 8.6 *Individual risk estimates compared*

societal risk estimates are against early death; the Canvey 2 data is for 'serious casualties'; the Sizewell 'B' data refers to early deaths and to fatal cancers. The latter data shows the predicted consequence of containment failure in the frequency region 10^{-7} and 10^{-8}, where the number of deaths increases markedly. The Sizewell data sets also exhibit a low individual risk relative to the

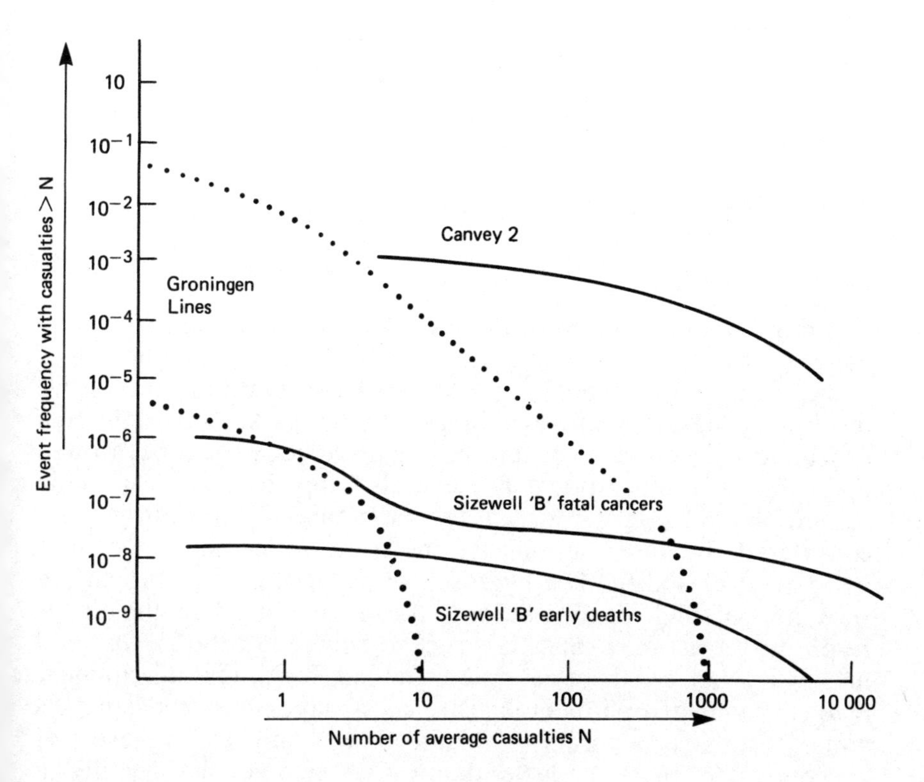

Figure 8.7 *Societal risk estimates compared*

societal risk when compared to the other sets. This is due to the wide spread of the random radioactive damage based on the hypothesis of linear extrapolation.

Toxicity relationships

Comments have already been made on toxicity in Chapters 1 and 2. Whereas the inadvertent release of toxic material in powder or liquid form can have very serious consequences (as at Seveso), it is the release of dangerous gas (as at Potchefstroom or Bhopal) which is much more likely and more difficult to estimate.

The factors which affect the estimating of toxic gas hazads have been reviewed by a number of authors and include:

1 toxic mechanisms;
2 uncertainty of experimental data;
3 inhalation rates;
4 vulnerable people;
5 shelter.

Each of these will now be discussed in turn.

Toxic mechanisms

The most common hazardous gases in the chemical industry act as irritants. In this context irritation is a technical term: the effect ranges from mild discomfort to death. An irritant gas attacks the respiratory passages and the lungs. For gases such as ammonia which are very soluble in water, the damage tends to be concentrated upon the respiratory tract. The eye also appears to be especially vulnerable as a target organ for ammonia – eye injuries are considered the most serious hazard from ammonia in terms of permanent disability. The hygroscopic properties of ammonia are such that in high concentrations the initial impact in the upper respiratory tract may cause obstructive oedema of the larynx, and in severely affected cases speedy action by a multidisciplinary medical team of ophthalmologists, chest physicians, anaesthetists and surgeons is necessary.[19] On the other hand, at low levels of concentration there is little doubt that people can acquire a tolerance towards ammonia so that a level which is obnoxious or even intolerable to some people may pass unnoticed by others who work habitually in such levels.

This is not the case with less soluble gases such as chlorine, which acts chiefly upon the lungs and whose injurious effects are due to its strongly oxidative properties. The extent of injury from chlorine depends upon the concentration of the gas, the duration of the inhalation period and the susceptibility of the exposed individual. In severe cases cellular damage to lung tissue and alveoli, due to the action of elemental chlorine, hypochlorous acid and hydrochloric acid, may after several hours lead to a chemical tracheo-bronchitis and pulmonary oedema. If secondary infection occurs, broncho-pneumonia may develop after a few days.[20]

In cases where exposure to chlorine has been severe, admission to hospital is necessary and after the appropriate remedial action has been taken surveillance of the patient should be maintained for a prolonged period. With such care, after recovery from exposure to chlorine no residual disability is expected to result.

Of the other industrial gases listed in Table 2.2, bromine, which is more soluble than chlorine but less soluble than ammonia, attacks both the respiratory tract and the lungs.

Hydrogen sulphide is an irritant gas but also attacks the nervous system and causes respiratory paralysis. It is oxidized in the bloodstream to pharmacologically inert compounds.

Hydrogen fluoride is an irritant gas but also gives rise to fluoride poisoning in the body.

Phosgene is an extremely unpleasant and deadly gas, producing little irritation in the upper respiratory tract but causing severe and permanent damage to the lungs including lung oedema and pneumonia.

Methyl isocyanate is even more deadly than phosgene, and is both strongly irritant and systemic in its action.

Hydrogen cyanide is the only one of the gases listed in Table 2.2 which is not an irritant; it causes cyanide poisoning. The most important effect of this is probably the inhibition of cytochrome oxidase, which in turn prevents the utilization of molecular oxygen by the cells. The cyanide is excreted in the urine.

Uncertainty of experimental data

The primary source of information on the lethal toxicity of industrial gases is from experiments with animals, particularly mice. As was remarked in Chapter 1, determining inhalation toxicity values from animal experiments is a difficult undertaking. The breathing rate of the animal and the caging conditions can affect the result, as can such factors as the individual susceptibility of the animal, its health, its age and the methodology of the experiment. An early experimenter, Professor Lehmann, provides a cautionary account of two apparently similar dogs being subjected to the same treatment with hydrogen sulphide.[20] One dog became ill after an hour and died nine hours later, the other dog came out of the gas chamber after five hours' immersion with no apparent ill-effect whatever. To obtain an average value with a high level of confidence, a sufficient number of animals needs to be employed. In addition to the animals being restricted to one species, they need to be selected on the basis of health, age, bodyweight and possibly sex.

The toxicity value most widely used and sought is the concentration value at which, for a given exposure time, the mortality in a group of animals is 50 per cent. This value is called the LC_{50}.

The pioneering work of Trevan was also mentioned in Chapter 1. He showed that for a particular dose–mortality determination, the confidence level depends both on the number of animals and on the mortality. For a given confidence level it is necessary to use more animals to determine an LC_{10} or LC_{90} than to determine an LC_{50}. Alternatively, and this is the more usual experimental case, for a given number of animals in the sets subjected to various concentrations at each expected level of mortality, the confidence in the LC_{10} and LC_{90} values is less than that in the LC_{50}. Table 1.3 reflected this situation with data derived from experiments upon dogs subjected to various concentrations of chlorine. The confidence levels are likely to be very low for the LC_{01} and LC_{99}.

The available toxicity information varies widely from one gas to another. For some gases there may be but a single reported set of experiments on one species at one exposure time and with little distinction between immediate and delayed deaths. (Animals may not die immediately and it is necessary to observe delayed deaths over a number of days.) For others there may be several sets at differing exposure times, and for yet others there may be several sets on differing species.

There is an appreciable variation in the values given by different workers for the same species. Thus, for chlorine there is a factor of about 2 in the LC_{50} values reported for mice for 30 minutes' exposure.

There are major differences between the species in their respiratory anatomy and physiology.[21] It is therefore not surprising to find significant differences reported in the LC_{50} for different species, and the extrapolation of animal values to man is often a matter for debate. This was illustrated in Figure 1.4, and Tables 8.7 and 8.8 enumerate some of the underlying anatomical differences. The data in Table 8.8 may be derived from that in Table 8.7.

Taking all the factors into account, it appears reasonable to assume that the animal data on the acute effects of the dangerous industrial gases has a broad correspondence to man. There is no reason to suppose that the small animals are less sensitive than man, indeed the literature suggests that where chlorine, ammonia and hydrogen cyanide are concerned the larger animals are relatively less susceptible than smaller ones. The LC_{50} values given in Table 2.2 are averaged over a range of species and, if a body-weight factor does exist, are somewhat pessimistic.

Table 8.7
Comparative anatomical data

Species	Bodyweight (kg)	Lung volume (ml)	Alveolar surface (sq. m)	Minute volume at rest (ml/min)
Human	75	7000	82	6000
Dog	22.8	1501	90	2923
Monkey	3.7	184	13	694
Rabbit	3.6	79	5.9	1042
Rat	0.14	6.3	0.39	84
Mouse	0.023	0.74	0.068	24

Table 8.8
Comparative respiratory data

Species	Minute volume per bodyweight (ml/min/kg)	Minute volume per lung volume (ml/min/ml)	Minute volume per alveolar surface (ml/min/ sq. m)	Lung volume per alveolar surface (ml/sq. m)
Human	80	0.86	73.2	85.4
Dog	128	1.95	32.5	16.7
Monkey	188	3.77	53.4	14.2
Rabbit	289	13.2	176.6	13.4
Rat	600	13.3	215.4	16.2
Mouse	1043	32.4	352.9	10.9

The lethal concentration is a function of concentration and time and is generally considered to relate to the load equation given in Chapter 1:

$$C \times T^n = \text{constant}$$

For many of the irritant gases n seems to approximate to 0.5, but the experimental data from which this parameter is derived is very sparse in most cases and does not always agree. Work by Doe and Milburn[22] gives a value of $n = 1$ for some other gases. The values given in Table 2.2 are either 0.5 or 1 in recognition of the approximate nature of the data from which they have been derived.

Inhalation rates

Some allowance needs to be made for breathing rate when applying to the LC_{50} values derived from animal experiments to human activity.

Information on the breathing of animals during exposure is recorded in some experiments; Lehmann's pioneering work always included quantitative information, but this has rarely been repeated. It has been observed that some animals such as dogs and rats tend to become very passive and have slow breathing rates during exposure; others such as cats and mice tend to become excited.

By contrast, the variation of man's inhalation rate with differing degrees of exercise is well documented. It is not easy to apply this information, however. At one extreme some members of the population may deliberately minimize their physical activity, perhaps by staying indoors as a result of advice from an emergency plan, while others may enhance their activity levels, perhaps by running through the gas cloud in an attempt to escape.

Ideally some form of toxicokinetic model should be used,[23] but for the irritant gases a simple approach is to define the physical activity in terms of the breathing rate and to assume that there is a linear relationship between this and the allowable gas absorption. Thus, at twice the activity half the concentration is allowable.

Vulnerable people

Extrapolation from animals to man is in the first instance usually done for healthy young adults. It is necessary, however, to consider the effect upon more vulnerable members of the population, such as the very young and people with respiratory problems. The literature has little information on this aspect, although it is important because a large part of a toxic gas cloud could be at concentrations where the vulnerable elements of the population form most of the casualties.

An estimate of the relationship between the lethal concentrations of chlorine and ammonia at varying mortalities for a general and a more vulnerable population has been given by Eisenberg *et al.*[24] An alternative approach is to start with the minimum concentration at which a fatality has been recorded in any of the animal experiments, and to use a concentration/mortality slope of the same magnitude as that obtained from the animal tests.

The effect of shelter

An effective safeguard against even rapid and large releases at short distances is to remain indoors, or to go indoors quickly and keep doors and windows shut until the cloud has passed. Thus, following the instantaneous release of nearly 20 tonnes of anhydrous ammonia at Houston, Texas, in 1976, it was observed that many people caught nearby in the open were killed or seriously affected, while people in an adjacent office block were mostly unharmed.

The ratio of inhaled dose inside a shelter to that outside is well documented with information on such influences as the outside wind speed, the number and type of openings, and so on. A reasonable average figure for a modern house in the UK is a factor of 3 for exposure times up to an hour.[25, 26]

Table 8.9 provides a summary of the contrasting lethal concentrations of chlorine taking into account the various factors that have been reviewed (for a more detailed explanation of its derivation, see Withers and Lees).[27]

Table 8.9
Lethal concentrations of chlorine (ppm) at an exposure period of 30 minutes

	Mortality 10%	50%	90%
Healthy people outdoors			
up and about	125	250	500
vigorous walking	62	125	250
Healthy people indoors			
in bed	750	1 500	3 000
up and about	375	750	1 500
Vulnerable people outdoors			
up and about	50	100	200
vigorous walking	25	50	100
Vulnerable people indoors			
in bed	300	600	1 200
up and about	150	300	600

References

1 Baker, W. E., Cox, P. A., Westine, P. S., Kulesz, J. J. and Strehlow, R. A., *Explosion Hazards and Evaluation*, Amsterdam: Elsevier, 1983.

2 Hopkinson, B., British Ordnance Board Minutes 13565, 1915.
3 Zuckerman, S., 'Experimental study of blast injuries to lungs', *The Lancet*, p. 219, August 1940.
4 Baker *et al.*, op. cit. note 1, Figure 2-3.
5 Irving, D., *The Destruction of Dresden*, London: William Kimber, 1983.
6 Bowen, I. G., Fletcher, E. R. and Richmond, D. R., *Estimate of man's tolerance to the direct effect of air blast*, US DASA 2113, Washington, DC: Lovelace Foundation, October 1968; White, C. S. *et al., The biodynamics of air blast*, US DASA 2738T, Washington, DC: Lovelace Foundation, July 1971. (Note that these are but a sample.)
7 Glasstone, S. and Dolan, P. J., *The Effects of Nuclear Weapons*, London: Castle House Publications, 1980.
8 'Notes on the basis of outside safety distances for explosives involving the risk of mass explosion', presented to UK Interdepartmental Explosives Storage and Transport Committee, 1959. Summarized by Jarrett, D. E., in 'Derivation of British explosives safety distance', *Annals of the New York Academy of Sciences*, 152, p. 18, 1968.
9 Withers, R. M. J. and Lees, F. P., 'The lethal effects of a condensed phase explosion in a built-up area', LUT Report MHC/86/3, *Journal of Hazardous Materials*, 15, 1987 (under review).
10 TNO, *Analysis of LPG incident, Mexico City, November 1984*, Report 8727–13325, Ref. 85–0222, 1985.
11 Scilly, N. F. *et al., The effects of explosions in the process industries*, Report of Overpressure Working Party to Major Hazards Assessment Panel, Rugby: I. Chem. E., 1985.
12 Davenport, J. A., 'A study of vapour cloud incidents', *I. Chem. E. Symposium Series 80*, 1, c1, 1983.
13 *Canvey Second Report*, London: HMSO, 1981.
14 Ibid.
15 TNO Report, op. cit. note 10.
16 *Biological effects of ionizing radiations*, BEIR III Report, National Research Council's Advisory Committee, Washington, DC: National Academy Press, 1980.
17 Blix, H., 'ENC1986 Report', *Atom*, 358, p. 2, 1986.
18 Gittus, J., 'Risk assessment for the PWR', *Atom*, 352, p. 7, 1986.
19 Withers, R. M. J., '*First Report of MHAP Toxicity Panel*',

Proceedings of International Chlorine Symposium, London 1985, Chichester: Ellis Horwood, 1986.
20 Lehmann, K. B., 'Experimental studies on acquisition of tolerance to industrial gases (ammonia, chlorine, hydrogen sulphide)', *Archiv für Hygiene*, 34, p. 272, 1899.
21 Altman, P. L. and Dittmer, D. S., *Biology Data Book*, Vol. 3, Bethesda: Fed. Am. Soc. for Exp. Biol., 1974.
22 Doe, E. and Milburn, G. M., *The relationship between exposure concentration, duration of exposure and LC_{50} values*, Guildford, British Toxicology Society, March 1983.
23 Withers, R. M. J. and Lees, F. P., 'The factors affecting lethal toxicity estimates and the associated uncertainties', *Hazards IX*, Rugby: I. Chem. E., 1986.
24 Eisenberg, N. A., Lynch, C. J. and Breeding, R. J., *Vulnerability model*, National Technology Information Service (US), Report AF–A105–245, Springfield, Va., 1975.
25 Davies, P. C. and Purdy, G., 'Toxic gas risk assessments – the effects of being indoors', *I. Chem. E. NW Branch Symposium Papers No. 1,* 1.1, 1986.
26 *Reactor Safety Study*, App. VI, Section II WASH–1400, Washington, DC: US Regulatory Nuclear Commission, October 1973.
27 Withers, R. M. J. and Lees, F. P., 'The assessment of major hazards: the lethal toxicity of chlorine: Part 2, Model of toxicity to man', *Journal of Hazardous Materials*, 12, p. 283, 1985.

9 Assessing the impact upon a local population

If the assessment of a major hazard is to include a determination of the societal risk, the likely whereabouts of the people at risk need to be known. They are in two groups: those at work on site, and those living and/or working in the vicinity.

There should be no difficulty in securing the necessary information on the first of these; it should be available as a routine part of the on-site emergency plan procedure. However, it is not at all straightforward in respect of the off-site population. Aspects which need to be known include:

1 location and numbers of people normally resident at night;
2 daytime variation to this data;
3 the numbers and location of more vulnerable people;
4 proportion of people outdoors.

In Chapter 1 the various phases of a risk assessment were emphasized. The accuracy with which the population characteristics need to be described depends upon the requirements and results of other phases in the assessment. The methodology employed is an iterative one and should only call for detail and extra precision as and where these are really necessary.

Methodologies of risk estimates

A number of methods by which risks to the public have been estimated have been described in the literature.

A common initial procedure is to plot on a map the contours of the physical effect (thermal radiation, overpressure, toxic load) from a single hazardous event, and then to convert these into individual risk contours using probit equations. In Chapter 1 the limitations to the accuracy of probits were explained and the large relative errors that are possible outside the 20–80 per cent mortality range made clear. There is also some debate about the appropriate constants to be used in the probit equations, depending upon the choice of primary data and the degree of conservatism applied. Nevertheless, it is possible to tabulate this individual risk data into segmented compartments. These segments are commonly based upon either 30° or 45° sectors radiating from the hazard source. Societal risks can then be obtained from the individual risk tabulations if the population density within each segment is known. Table 9.1 gives some illustrative data on the probit constants that may be employed.

In the first Canvey report[7] the assessment was confined to selected residential areas where a uniform population density of 4000 persons per square kilometre was assumed. However, in Appendix 3 to the report Beattie describes an investigation into chlorine toxicity where he assumed uniform densities per square kilometre of 5000 persons in the residential areas and 100 persons

Table 9.1
Illustrative probit parameters

(probit = $k_1 + k_2 \ln x$)

Phenomenon	Source	x	Parameters k_1	k_2
Thermal radiation	Eisenberg[1]	$tI^{4/3}$	−14.9	2.56
Overpressure	Eisenberg[2]	p	−77.1	6.91
Ammonia inhalation	Eisenberg[3]	$C^{2.75}T$	−30.6	1.38
	MHAP Toxicity Panel[4]	C^2T	−35.9	1.85
Chlorine inhalation	Eisenberg[5]	$C^{2.75}T$	−17.1	1.69
	Withers and Lees[6]	C^2T	− 8.29	0.92

where C = concentration in ppm
I = radiation intensity (kilowatts per square metre)
p = peak overpressure (force per square metre)
t = time in seconds
T = time in minutes

in the rural areas. In Appendix 14 Fryer *et al.* present casualty estimates for a large ammonia release. From the national census data they obtained the number of people living in 100 metre squares around the hazard source, and then converted this into a segmental tabulation based on twelve 30° sectors. In this study the population profile was extended as far as 32 km from the hazard source, while there were few in number within 1.5 km from the source. These studies did not distinguish between day and night-time (the census data is for night-time) or consider the effects of vulnerable people or shelter.

In the second Canvey report[8] the method used by Fryer *et al.* seems to have been generally adopted.

The authors of the Rijnmond Report were much more thorough. In the first place they considered the workers on site. Numbers were obtained from the various managements, both for the working day (N_d) and for other times (N_n). Allowance was made for holidays (h weeks per year) and sickness, as well as for three-shift, seven-day working. Thus, the average number of people on site during the daytime was $(5N_d + 52 \times 16N_n)/[5 \times (52 - h)]$. The average density for the six large sites examined was found to be 200 persons per square kilometre.

For the off-site population, estimates were obtained from the national census of the number of people living in 500-metre squares around the sites. Subsequent cross-checks were made by counting the number of dwellings and multiplying by statistical data on the number of people in occupation per dwelling. Data was also obtained on the number of workers in each district and separate compilations were made of the day and night-time populations. The population grid covered an area of some 75 square kilometres. Outside this area a uniform population density was assumed but this was only applied in the very few cases where exceptionally large hazardous clouds were postulated. The average density inside the grid was about 600 persons per square kilometre.

The 500-metre squares were then subdivided into four 250 × 250-metre squares and the population for each of these assumed to be 25 per cent of the larger squares. Originally an attempt was made, using probit equations, to calculate the likely fatalities for each point of the population grid for every combination of likely event. This attempt had to be abandoned because of the constraints of available computer capacity, and a shorter procedure was adopted.

Event trees

The construction of an event tree to form the basis of a complete assessment of a major hazard is a relatively straightforward exercise in logic, but the numerical calculations which follow are so extensive and tedious that a computer is essential.

The main branches of the event tree may include:

1 a set of release quantities and their frequencies;
2 a set of on-site circumstances involving such factors as:
(a) chances of on-site ignition;
(b) chances of on-site explosion;
(c) duration and relevance of working periods, weekends, day, night, and so on;
(d) extent of shelter, evacuation;
3 a set of off-site locations, usually based on sectors;
4 a set of weather conditions, usually restricted to a representative set using the Pasquill categorizations and two wind speeds;
5 a set of chances for explosion and ignition and other off-site factors similar to those in 2 above, together with considerations for more vulnerable people;
6 a set of dispersion formulae and damage relationships of the type explained in previous chapters.

If the set of off-site locations were to be made up of twelve sectors with twenty positions in each sector, ten release quantities and five weather combinations would constitute over a thousand computer runs without consideration of some of the finer points listed above. If the numerical solution to the combined dispersion equations and damage relationship is too tedious, it is possible that the computer time for all the runs of the hazard assessment may exceed the likely duration of the hazard itself!

Some understanding of the magnitude of the problem may be gleaned from a brief examination of the kinds of equation associated with the conservation models for gas dispersion mentioned in Chapter 6.

Conservation models begin on the firm theoretical basis of the exact partial differential conservation equations for total mass, momentum, energy and species using mean time averaged quantities in a turbulent field. But the precise form of these equations may differ between the various authors. A typical set due to Ermak, Chan *et al.*[9] is as follows:

mass:

$$\nabla\ (\rho u) = 0$$

momentum:

$$\frac{\partial(\rho u)}{\partial t} + \rho u.\nabla u = -\nabla p + \nabla(\rho K^{m}.\nabla u) + (\rho - \rho_h)g$$

energy:

$$\frac{\partial\theta}{\partial t} + u.\nabla\theta = \nabla(K^{\theta}_{\cdot}\,\nabla\theta) + \frac{C_{pn} - C_{pa}}{C_p}(K^{\omega}_{\cdot}\,\nabla\omega).\nabla\theta + S$$

and NG vapour (species equation)

$$\frac{\partial\omega}{\partial t} + u.\nabla\omega = \nabla.(K^{w}.\nabla\omega)$$

Also required are ideal gas law equations for density

$$\rho = \frac{M_n M_a P}{RT(M_n + (M_a - M_n)\omega)} = \frac{MP}{RT}$$

and for specific heat

$$C_p = C_{pa}\ (1 - \omega) + C_{pn}\omega$$

where u = (u, v, w) is the velocity;
ρ = density of the cloud mixture;
g = acceleration due to gravity;
θ = potential temperature deviation from an adiabatic atmosphere;
S = temperature source term (for example latent heat);
ω = mass fraction of NG vapour;
P = absolute pressure;
R = universal gas constant;
T = absolute temperature;
M = molecular weight of the mixture;

K^m, K^θ, K^ω are diagonal eddy diffusion tensors for momentum, energy and NG vapour;
n, a are subscripts denoting NG and air.

Such equations require large computational facilities for their solution even though recourse to considerable approximations and simplifications is made.

Although the conservation models have a number of potential advantages (as explained in Chapter 6), where the prediction of downwind range under average conditions is concerned the results are not so different from those of the simpler 'box' models (as illustrated in Table 6.4). Such a 'box' model was used by the authors of the Rijnmond Report,[10] but nevertheless they had to resort to an approximate methodology in their computation of societal risk. This was to determine the radius at which the probability of death was 50 per cent and then to assume that the number within this radius who escape was balanced by the number outside who would be killed.

Rapid methods of hazard assessment

The experience of hazard assessment work of the kind just described suggests that there is a requirement for a rapid methodology. Even if it is only to be used in an initial screening of the situation to determine those locations or those events which require more careful study, much valuable time may be saved.

The main steps in a simplified procedure as applied to the estimation of fatalities in a fire and explosion scenario were outlined in Chapter 7. This rapid methodology was based upon the following four prime components:

1. the use of standard population densities as explained in the following section of this chapter;
2. a 'standard' release equation as explained in Chapter 5, initially keyed to the magnitude and frequency of the 'top' event;
3. a simple scaling law (described in Chapter 6) instead of the more fundamental equations given in the previous section of this chapter;
4. the use of an impact model which is based upon:
   (a) sectors of uniform population density;
   (b) an inverse power law for the decay of the intensity of the physical effect with distance;

(c) a lognormal distribution (for example a probit equation) for the relation between the causative factor and the probability of death.

The theoretical limitations of this impact model have been discussed in papers by Lees, Poblete and Simpson.[11] They have shown that in many cases the number of casualties will equal the number of people inside the 50 per cent contour, and that in others a simple correction factor can be derived to compensate for the error in the method. Thus, the model reduces to the approximate method used in the Rijnmond study under defined conditions.

In other words, while it may not always be true that on average as many people will survive inside the LC_{50} contour as die outside, there must be a contour where this does apply and the difference between this contour and the LC_{50} may not be great in many cases, although in others it will.

The basic equation given by Lees *et al.* is:

$$N = \pi R_{50}^2 DC$$

where N = number of casualties;
R_{50} = radius for 50 per cent probability of death;
D = population density;
C = $\exp(2\sigma^2/n^2)$;
σ = the spread parameter of the lognormal distribution.

If C is set equal to unity the model reduces to the approximate method used in the Rijnmond study. The authors of that study tested their approximation by calculating more rigorously the numbers of people between an R_5 and an R_{50} and between an R_{95} and an R_{50}. Very small differences were found in the on-site case but agreement was not so close in residential areas. Nevertheless, the error was not likely to exceed 25 per cent in the general case. This is acceptable for an approximate method and significantly less than the likely error in other elements of the chain.

Studies made with the aid of simplified models (compare Chapter 7) have suggested that in most cases larger errors in the cumulative fatalities will arise from faulty location of population boundaries than from incorrect estimates of density. Moreover, the contributions from the far field are not so critical as are those from close in.

Population density

The first Canvey report was based upon the use of a standard population density for residential districts of 4000 people per square kilometre. One of the most well-known and comprehensive studies into population densities is Abercrombie's.[12]

Table 9.2, taken from his report, lists the population densities in built-up areas around pre-war London. It distinguishes between the average densities of the built-up housing estates and the grossed-up areas which include vacant land, sports fields, parks, and so on. The average of the latter is 15.7 persons per acre or 3877 per square kilometre. It is unlikely that the present-day position will be very different since factors such as smaller families will be compensated by factors such as less vacant land.

Thus, the average of 4000 persons per square kilometre used in the Canvey report seems well supported in so far as residential districts in the UK are concerned.

Table 9.2
Population data after Abercrombie

Place	Residential density (persons per acre)	Gross density (persons per acre)
Dunstable	21	10
Luton	20	19
Windsor	18	18
Slough	22	19
Hornchurch	19	19
Romford	22	19
Tilbury	20	20
St Albans	14	10
Watford	17	17
Dartford	19	17
Hayes	29	25
Hillingdon	15	13
Staines	16	14
Banstead	10	9
Coulsdon	14	14
Epsom	14	11
Guildford	15	14
Woking	14	14
		Total 282

Average: 15.7 or 3877 persons per square kilometre

However, very much higher densities do occur in many built-up areas in the UK and other countries where major hazards exist, and these higher densities need to be taken into account in any assessment of the risk to populations at distances less than 1 kilometre from the hazard centre. Nowadays such concentrations are usually confined within narrow boundaries, but in pre-war central London they were widespread. The prime purpose of the Abercrombie plan was to disperse these concentrations of people and associated industry outside the Greater London area altogether by the establishment of satellite towns.

Following the precepts of the Abercrombie reports the area of Greater London is best described under four headings:

1. the City of London itself;
2. the fifteen inner London boroughs;
3. the county of London;
4. parts of the Home Counties inside the Greater London area running into the green belt.

This whole area, bounded by the green belt, is surrounded by the remainder of the Home Counties. By 1943 London had experienced massive conventional air raids and a significant part of the population had been evacuated to the Home Counties and beyond. In many districts of postwar London the population density never rose again to the pre-war levels because of the planning measures that were adopted following the Abercrombie reports.

Table 9.3, again based on data taken from the Abercrombie reports, contrasts the 1938 and 1943 population distributions. This indicates that the figure of 600 persons per square kilometre suggested for rural areas may be conservative; indeed there are many countries in the UK with densities of 400 persons per square kilometre or less.

The average density of the inner London boroughs may seem high especially since they contain open spaces such as Battersea Park and Kensington Gardens where the density is almost zero. Places such as West Ham and Willesden, however, had extensive areas with around 30 000 people per square kilometre. Its said that in spite of Abercrombie and the new towns they were safer and more sociable places in many ways than they seem today.

When describing population patterns it is recommended that, as with other aspects of the risk assessment, a structured and phased approach be undertaken.

Table 9.3
Population density contrasts

Zone	Area (square miles)	Population 1938	Density (persons per square kilometre)	Population 1943	Density (persons per square kilometre)
1	1	9 000	3 476	4 200	1 622
2	28	1 585 000	21 815	808 000	11 121
3	90	2 468 000	7 548	1 395 000	4 266
4	2 480	2 188 000	243	1 207 000	143
Total	2 599	6 250 000		3 414 200	
Averages			662		361

In the first phase of a risk estimate, individual risks should be calculated on the basis of eight 45° sectors. This will allow consideration to be given to the variable weather pattern but does not involve population data. For gas clouds an eight-sector grid is generally sufficient, since the cloud tends to spread out at an angle of at least 45°. Further work then needs to be carried out to determine societal risk in those areas where the risk exceeds one in a million, and in some circumstances a twelve-sector grid may be required.

For distances greater than 1 kilometre from the source only two categories of population density need be considered:

1 4000 persons per square kilometre in defined residential areas;
2 600 persons per square kilometre outside these areas provided the density is obviously not zero (in estuarial waters, for example).

For distances greater than 0.4 kilometres from the source, the population boundaries should be estimated within 0.025 kilometres using a map with a scale factor of at least 1:10 000. Four typical densities may be used:

1 15 000 persons per square kilometre where there are exceptionally close back-to-back terraces or high-rise flats: these should be checked by making inquiries locally;
2 10 000 persons per square kilometre will apply wherever the

map indicates suburban housing of conventional layout;

3 1000 persons per square kilometre where there are widely spaced detached residential districts;

4 600 persons per square kilometre outside these areas provided the density is obviously not zero (in parks, playing fields, and so on).

For distances less than 0.40 kilometres from the source, data taken directly from the national census must be used together with maps having a scale of 1:2500. Because this data may be out of date it should be supplemented by inquiries made locally.

In the UK there is a full census every decade, the most recent being in 1971 and 1981. The results are published together with user guides. The basic small geographical unit for the census is the enumeration district (ED), which is an area of land defined in terms of the number of households which constitute a practicable workload for an enumerator. The average population of an ED is about 500 in urban areas and 150 in rural areas.

Initial calculations of societal risk may be based on the night-time census data particularly where the risk from continuous operation of process plant over 24 hours is to be assessed. Some hazards, for example those concerned with transport operations, can occur only in the daytime, however, when the numbers of people in the residential areas are significantly less. The daytime population may be estimated from the census night-time figures as illustrated in the next section of this chapter.

The estimation method described above requires the use of Ordnance Survey maps. Four principal map sizes are the 1:25 000, 1:10 000, 1:2500 and 1:1250 scale series. The 1:10 000 scale is that normally used by local authorities. It is also the size on which the EDs are marked out and made available by the Census Office. The 1:2500 and 1:1250 maps provide more accurate location of buildings but the whole country is not covered by these series.

Three illustrative examples of the method are now given.

Example 1 Cheshire

An area scan around a hazard source using 1:2500 maps suggests:

– 2.1 square kilometres with 15 000 per square kilometre giving 31 500 people

– 3.5 square kilometres with 10 000 per square kilometre giving 35 000 people

– 3.3 square kilometres with 1000 per square kilometre giving 3300 people

Total 7.9 square kilometres 69 800 people
10.1 square kilometres void

A population count was carried out using a sample of twelve EDs which taken as a group appeared to be representative of the whole area around the source. The number of people in this sample was then scaled up by the ratio of the total area to the sample area. The estimate of the population obtained in this way was 75 996, giving densities of 8539 persons per square kilometre in the populated areas and 3998 overall. This compares with an actual census value of 82 522 for this area. The percentage difference between the map-based estimate and the actual census value is 15 per cent, and that between the estimate based on ED ratios and the actual census value is 8 per cent.

Example 2 Essex

An area scan using 1:10 000 maps suggests:

– 0.56 square kilometres with 10 000 per square kilometre giving 5600 people
– 0.85 square kilometres with 1000 per square kilometre giving 850 people

Total 1.41 square kilometres 6450 people
1.41 square kilometres void

The ED count, based on a sample of ten EDs, gave a night-time population of 5446. This suggests a density of 3862 persons per square kilometre in the populated areas and a density of 1931 overall. This compares with the actual census value for the district of 1950 persons per square kilometre. The discrepancy in this case is one per cent, and for the map-based estimate it is 18 per cent.

Example 3 West Midlands

An area scan using 1:25 000 maps suggests:

– 0.69 square kilometres with 15 000 per square kilometre giving 10 350 people
– 2.44 square kilometres with 10 000 per square kilometre giving 24 400 people

– 0.55 square kilometres with 1000 per square kilometre giving 550 people

Total	3.68 square kilometres	35 300 people
	5.32 square kilometres	void

In this case no count is made of census EDs. The actual census value for the population density of the area was found to be 3598 persons per square kilometre and the difference between the map-based estimate of 3922 persons per square kilometre and the actual census is 9 per cent.

Population composition

The method just described gives an estimate based on the night-time population. An approximate estimate of the number of people over the whole 24 hours is 80–85 per cent of this value. However, the probability of some hazards may be a function of time of day; it may be necessary to distinguish between the risks in and out of doors; more vulnerable elements of the population may need to be taken into account.

For districts in the area covered by the Canvey Report the census enumeration district data in Table 9.4 applies.

There is clearly an appreciable variation in the numbers from one enumeration district to another. This emphasizes the importance of the original data when seeking a true density at distances close to the hazard source. The census data may be out of date and a fresh local inquiry may have to be undertaken.

Table 9.5 is based on wider studies of census data and gives a typical breakdown of the numbers in the various categories of an average UK household. This suggests that when allowance is made for sickness, unemployment, and so on, the daytime population may on average be around half the night-time population. However, for a duration period of a whole week the proportion of the time when people are at work is very small. Thus, for the Rijnmond study, Cremer and Warner estimated only 45 hours out of a total 168 hours. This gives a weighted average population over the 24 hours of 82 per cent of the night-time figure and with a night-time home occupancy of approximately three persons. The 'vulnerable' categories account for approximately 25 per cent of the household.[13]

In contrast to the residential areas, far fewer factory people will be on site during the night-time than during the day. It is

Table 9.4
Canvey enumeration district data

Enumeration district	Night-time population	Daytime no.	Daytime %	Vulnerable no.	Vulnerable %
AD03	430	130	30	64	15
AD16	500	221	44	106	21
AD18	711	225	32	109	15
AD21	305	138	45	76	25
AL06	627	194	30	65	10
AL17	556	220	40	107	19
AG01	369	175	47	84	23
AG03	659	239	36	119	18
AN06	790	448	57	103	13
AN09	920	503	55	109	12
Mean % of night time			42		17

(Daytime population is night-time minus those at work and those of school age. Vulnerable is those over sixty-five and under five.)

Table 9.5
A household composition model

Total night-time household	2.71
Number who should be away at work	1.29
Unemployed	0.07
Allowance for sickness, etc.	0.12
Number actually away at work	1.10
Children and students	0.70
Children under five or ill	0.34
Children at school and college	0.35
Retired and disabled people	0.38
'Housepersons'	0.35
'Housepersons' at home in daytime	0.28

therefore essential to obtain relevant data directly from the factory management.

It is well known that the very young and the elderly are more vulnerable to certain types of hazard than the average fit and healthy person. This aspect is difficult to quantify although Table 9.6 based on information available at the time of the 1981 census provides some indication of the scale of the problem.

Table 9.6
Vulnerability data

Category	0–4 years	Over 65 years	Vulnerable	Non-vulnerable
At home	3 815 000	7 657 000	11 472 000	42 953 000
Deaths from 'domestic' accidents	601	5 773	6 374	4 212
Percentage			0.06	0.01
Deaths from pneumonia	920	45 753	46 673	4 422
Percentage			0.40	0.01

Hence the 'vulnerable' categories are six times more vulnerable than the 'non-vulnerable' to 'domestic' accidents. (Such accidents account for 71 per cent of all non-transport fatal accidents.) They are forty times more vulnerable to pneumonia.

There is very little specific information on toxicity in relation to vulnerable people although a paper by the MITRE Corporation[14] has shown that people over forty-five are more susceptible to the inhalation of methylene chloride. In the Eisenberg model, mentioned in Chapter 8, it was estimated that for toxic deaths 50 per cent of the vulnerable would die where only 3 per cent of the non-vulnerable would die. A different precept (based on the lowest known lethal concentration to any species) was also referred to in Chapter 8.

While on average the 'vulnerable' categories may comprise 25 per cent of the total population, it is to be expected that the people employed on the factory sites would be in a 'non-vulnerable' category.

In the Rijnmond Report assessment of a toxic gas hazard, the proportion of the population reckoned to be indoors was 99 per cent allowing for the fact that some people would seek shelter. A recent study of V2 incidents in London (which took place in 1945)[15] gives a figure of 98 per cent and this provides an indication of the number likely to be indoors in a built-up area if no warning is given. However, this may be an upper limit due to the wartime conditions and because no account was taken in the calculations of the people who were outdoors and escaped unhurt. An alternative model by the same authors gives a figure of just over 96 per cent.

Discussion

Three common ways of presenting risk assessments that have been described are by:

1 drawing contours of equal individual risk;
2 tabulating societal deaths per annum;
3 displaying frequency/fatality magnitude graphs (*f*/*n* curves).

The first of these methods works through the long chain of events covering release of the hazardous material through to the chance of death occurring to an individual permanently located at a point in space away from the source of release.

For a set of events the chances of death from each event are simply added together to give a total frequency. Such a set may arise from a series of releases of varying size and frequency from a single source, from a set of different sources, or a combination of the two. Such release data may result from direct observation, may be synthetic or conceptual, or a combination of both.

In practice the method usually begins by calculating the chance of death at various points at known distances from the source to give a frequency/distance relationship for each one of a number of sectors around the source. The differing meteorological conditions will yield a different relationship for each sector. The sectors usually number twelve or eight to make a total span of 360 degrees. The points in the different sectors with common frequencies may be joined together to form a risk contour. Clearly the contour may be more readily smoothed if it is based on twelve sectors, but there is less computation in working with eight.

Differing chances of death will apply as a result of the person being in or out of doors; the weather will have different effects in daytime from those at night; different persons have differing vulnerabilities. Thus, differing contours can be produced, for example for indoors and outdoors, for night and day, and so on.

One advantage of the use of risk contours is that the risk at a point can be compared directly with other risks, for example being killed by lightning. It may also be compared with 'acceptable' risk criteria, for instance the 0.1 chance in 10 000 given in the Royal Society's 1983 review.[16] It is common to plot a family of contours differing from each other by a factor of ten, starting at the 'acceptable' figure of 0.1 in 10 000 and showing an overall reduction of a factor of 1000.

The risk contours do not require any demographic data and their presentation can therefore be independent of maps showing population densities. Indeed the concept of standardized release data might be used to develop standard risk contours for various common hazards using averaged meteorological data applicable to the UK as a whole, for example. But the projection of such contours on a map showing the extent of populated areas gives an immediate indication of the risks to be run by people and where the consequences may be most severe. Such a projection of a risk contour will not, however, readily provide a direct estimate of the numbers of people at risk.

Whereas in earlier studies (for instance the first Canvey report) hazards were often thought to extend as far as 10 kilometres from a potential source, it is now believed that the hazard decays very rapidly with distance and a precise knowledge of population boundaries beyond 2 kilometres is not needed for the majority of possible major industrial hazards. The chances of death at a particular point can be obtained in at least two ways:

1 By summing the chances obtained from each release using a damage relationship. In the case of a toxic release this relationship might be expressed as a 'probit' curve which, for example, relates concentration to percentage fatality. Thus, for each release a concentration at each point would be calculated using dispersion formulae and then the percentage fatality obtained from the probit.
2 By calculating whether or not the position of the point is inside or outside the 50 per cent lethality contour. If the point is outside no chance of death is assumed, if inside then death is assumed. As the number of releases in the set is increased this method tends to give the same result as before but it is obviously in error for small samples of release sets. On the other hand, it is computationally easy and has the advantage of using only the R_{50} from the damage relationship. This R_{50} is always known with considerably greater confidence than other values, particularly the R_{10} and below. For large release sets the principal assumption of the method is that the number of people who are killed outside the R_{50} contour is balanced by the number of people inside who survive. The mathematical limitations of this method have been referred to in a previous section of this chapter.

For purposes of emergency planning an estimate is required of the number of people likely to be affected, and to this end methods of evaluating societal risk have been developed. These require information about the likely distribution of people around the hazard in time and space.

The concept of societal risk provides a measure of the chances of a number of people being affected by a single event. For purposes of risk assessment it is simplest and least ambiguous to measure deaths rather than casualties. An estimate of total casualties can often be derived from the projected fatalities, using the statistical data made available from case histories.

The chances reduce as the number of casualties increases. In the Canvey Report the probabilities were given for casualties exceeding 10, 1500, 3000, 4500, 6000, 12 000 and 18 000.

In the second Canvey report (which may be considered as a revision of the first) the probabilities given originally were seen to be unduly pessimistic and were significantly reduced. The data reproduced in Figures 1.5 and 8.7 are from the second Canvey report. Even so, present understandings may see these figures as pessimistic still. Thus, casualties exceeding 18 000 would now be seen as so infrequent as to be incredible and not likely to be reported upon.

In the Canvey reports no attempt was made to measure the risk or the likely casualties on site. There is a very great difference between the risks run by those who work on site and those who live in residential districts off site.

In the Rijnmond Report a table of cumulative frequencies compared the likely annual rates of death between those on and off site, and was reproduced in Table 2.3.

These figures are supported by the evidence from case histories and emphasize that even though more people are to be found off site than on site the absolute number of expected fatalities within a given period is always likely to be much greater on site. Such on-site estimates can form an important part of a risk assessment since they can be compared with historical statistical data of on-site fatalities and provide a means of checking the realism of the assessment. A well-trained and well-informed workforce on site should be well aware of the relative risks they run and behave accordingly. Their attitudes and activities can provide people off site with their best reassurance on safety matters.

In the Canvey Report all the probabilities from the various events which lead to an expectation exceeding the given number of casualties were added together, but no attempt was made in the Canvey reports to combine the chances with the fatality numbers to give a single figure comparable with the cumulative figures quoted from the Rijnmond Report. In the opinion of the Canvey Report writers such a step would have implied an acceptance of a linear dependence of total harm upon the number of casualties. That the general public perceives lower-frequency larger numbers with more alarm than higher-frequency smaller numbers has already been remarked upon in Chapter 1.

For all practical purposes the 'unacceptable' Groningen boundary line of Figure 1.5 is the same as the ammonia plant *f/n* line given in the Rijnmond Report. It is clear that the result of multiplying the frequencies and numbers of the Canvey ammonia study would give a much more pessimistic societal risk than was found to be the case with the later Rijnmond ammonia study. Part of the explanation for this difference lies in the population data employed, but it should also be noted that the Rijnmond study used a toxicity relationship that would now be considered somewhat pessimistic.

References

1 Eisenberg, N. A., Lynch, C. J. and Breeding, R. J., *Vulnerability model*, Nat. Tec. Inf. Service, Report AF–A105–245, Springfield, Va., 1975.
2 Ibid.
3 Ibid.
4 Withers, R. M. J., 'Second Report of MHAP Toxicity Panel', *I. Chem. E. NW Branch Symposium Papers No. 1*, 6.1, 1986.
5 Eisenberg *et al.*, op. cit. note 1.
6 Withers, R. M. J. and Lees, F. P., 'The assessment of major hazards: the lethal toxicity of chlorine: Part 2, Model of toxicity to man', *Journal of Hazardous Materials*, 12, p. 283, 1985.
7 *Canvey First Report*, London: HMSO, 1978.
8 *Canvey Second Report*, London: HMSO, 1981.
9 Ermak, D. L., Chan, S. T., Morgan, D. L. and Morris, L. K., 'Dense gas dispersion model simulations', *Journal of Hazardous Materials*, 6, p. 129, 1982.
10 *Rijnmond Report*, Dordrecht, Netherlands: Reidel, 1982.
11 Poblete, B. R., Lees, F. P. and Simpson, G. B., 'Estimation of injury and damage around a hazard source using an impact model', *Journal of Hazardous Materials*, 9, p. 355, 1984; Lees, F. P., Poblete, B. R. and Simpson, G. B., 'Generalization of

the impact model for the estimation of injury and damage', *Journal of Hazardous Materials*, 15, 1986.

12 Forshaw, J. and Abercrombie, P., *County of London Plan*, London: Macmillan, 1944; Abercrombie, P., *Greater London Plan*, London: HMSO, 1945.

13 Petts, J. I., Withers, R. M. J. and Lees, F. P., 'The assessment of major hazards: the density and other characteristics of the exposed population around a hazard source', *Journal of Hazardous Materials*, 1987.

14 Huston, J. M., *Development of a Methodology to Identify Susceptible Human Populations*, USA: MITRE Corpn, 79W, July 1980.

15 Petts *et al.*, op. cit. note 13.

16 The Royal Society, *Risk Assessment*, Study Group Report, p. 180, 1983.

10 Transport risks

The explanations of methods for estimating risks have so far been confined to fixed installations. However, the records of case histories of accidents involving hazardous industrial materials reveal that comparable numbers of fatalities result from transport activities.

Thus, Kletz's list of major hazards[1] over the ten years 1970–79 gives 1474 people killed worldwide at fixed oil and chemicals installations while 1012 people were killed in connection with the transport of oil and chemicals. The average number of transport incidents killing five or more people is given as four per year, with a range of two to nine; the number of people being killed per incident ranged from five to 211. The worst incident was the Spanish road tanker tragedy described in Chapter 2. This occurred in the worst year (1978) when 384 people were killed, the best year being 1970 when only twelve people died. Only one of all these transport incidents occurred in the UK, in 1970, when a tanker in the Manchester Ship Canal overfilled with petrol and the spillage ignited, killing six people.

Whilst volumes of statistical information which may relate to transport accidents are made available to the public every year, the quantification of events which can lead to a major hazard lacks the degree of certainty which characterizes some of the fault trees constructed for fixed installations. The causes of transport hazards seem to be as various as they are rare, but human errors often play a large part.

A very large proportion of the accidents listed by Kletz happened to sea-going tankers carrying petroleum products. Transport accident statistics are often presented on a comparative basis in terms of accidents per mile travelled, or perhaps per tonne mile or passenger mile. Such comparisons suggest that sea transport is by far the safest, as Table 10.1 illustrates.[2] But ships do not appear to be anything like so safe in port. Many accidents occur in harbour manoeuvres and transfers, the accident rate being reckoned at 3.6 per thousand movements.

Obviously such tables of 'apples and pears' presuppose that the accidents are strictly comparable if the data comparison is to be meaningful. This is unlikely. Neither can it be supposed that an accident to a 100 000-tonne supertanker in harbour will necessarily be more catastrophic than to a 20-tonne road tanker in a country road, as the Spanish tragedy described in Chapter 2 makes clear.

Attempts can also be made to compare the risk of early death to an individual as a result of various transport modes. Table 10.2 is from Meslin,[3] who derived the French data for himself, but obtained the US data from the WASH–1400 study.[4] Here the ultimate accident criterion is strictly comparable, but some of the intermediate computations are not. Here the aim was to demonstrate that chlorine transport was a low-risk activity in France as well as the USA.

While there are various inferences to be drawn from national and international transport statistics, the main point is that the

Table 10.1
Comparative transport accident rates

Transport mode	Accidents per 100 million miles
Ships in open water	0.06
US train car	34
UK goods vehicles	79
UK private cars	110
US trains as a whole	1 500

(78% of US train accidents are derailments which do not affect all the cars but stop the train. US trains have a lot of cars but the track is in a poor condition relative to European systems.)

Table 10.2
Individual risk comparison

Probability of early death per annum Transport mode	France	USA
Private car	1.5×10^{-5}	3×10^{-4}
Railway	7.8×10^{-8}	4×10^{-6}
Air travel	1.2×10^{-7}	9×10^{-6}
Chlorine shipment by rail tanker	1.5×10^{-9}	5×10^{-9}

proportion of the total goods traffic which is made up from hazardous cargoes is quite small. In the UK road traffic predominates, as Table 10.3 makes clear.

According to official statistics there are around 220 000 road traffic accidents every year of which some 34 000 (14 per cent) occur to goods vehicles. There are over 1.5 million goods vehicles registered in the UK and they travel about 23 000 million miles in a year giving about 140 accidents in 100 million miles. Most of these vehicles are under 1.5 tonnes and only 1.1 per cent are designed to carry hazardous materials. About 0.3 per cent are designed to carry chemical products and 0.8 per cent to carry petroleum products, split fairly evenly between rigid and articulated vehicles. The 220 000 accidents are responsible for around 8000 fatalities every year, so on the basis of these statistics one might suppose that between five and one per cent of the 14 per cent × 8000 fatalities could be associated with road traffic accidents involving hazardous loads. These numbers, between 56 and 11, are based on the assumption that vehicles carrying hazardous loads are as accident-prone as other road users, and that they have the same

Table 10.3
UK hazardous traffic movements

	Millions of tonnes in 1974					
Transport mode	Petroleum products	% all	Other chemicals	% all	Total hazardous	Total all freight
Road	85	5.2	34	2.0	119	1707
Rail	19	10.5	3	1.7	22	176
Other	6	6.0	1	1.0	7	102
Totals	110	5.5	38	1.9	148	1985

accident:fatality ratio. Between 16 and 3 of them might be associated with chemical products.

The actual number of recorded deaths in the UK due to chemical loads is nineteen over the nine years 1968–76, with fifty to sixty serious injuries. Anderson[5] has reported that if the drivers of the vehicles involved in the accidents are excluded there are only two fatalities involving hazardous road transport over the eight years 1970–77. The Chemical Industries Association[6] reports that over a recent twelve-year period only one road traffic death has been caused by chemicals.

Thus, the estimate just obtained from general statistics seems to be rather pessimistic in the light of recent actual events. The discrepancy could be attributed to:

1 vehicles transporting hazardous material being less accident-prone than other vehicles;
2 accidents to hazardous vehicles causing fewer deaths than other road accidents.

Both these hypotheses are likely to be true since:

1 Being larger than the average vehicle there are fewer of them on the road than other vehicles for an equivalent tonnage transported. Moreover, statistics show that the accidents per mile are less for goods vehicles than for private cars so that they are likely to be lower still for hazardous transport whose drivers are specially trained to work to a code and whose vehicles are regularly inspected and maintained to a high standard.
2 Vehicles with hazardous loads are often routed away from built-up areas, the code would require more headway than is commonly practised, and the vehicles would not be carrying passengers.

When extrapolated to the circumstances of a single fleet operator the numbers become very small as the following example makes clear.

Hypothetical example of risks from hazardous road transport

Assume that a fleet of ten tankers operates out of a storage depot deliving hazardous petroleum products to a variety of customers.

In the light of the figures quoted previously, assume that each tanker averages 15 000 miles annually. Based on the previous discussion, assume that the fleet is well managed and that an average incident rate of 25 per 100 million miles applies.

The total annual mileage will be 150 000 and the total annual accidents will be .038 (or 38 in 1000 years). Remember that these are *average* figures and that there will be a wide scatter amongst these random events so that in some years the figures will be higher and some years lower. Remember also that these are *reportable* accidents – many more unreported accidents will happen to these vehicles.

The 1981 survey published by the UK Department of Transport established that the risk of fire to vehicles following road traffic accidents was one in 281 for all vehicles but only one in 2500 for goods vehicles.

If we assume that the chance of a fire following a traffic accident to one of the tankers is similar to the latter figure, the number of fires arising as a result of road traffic accidents during fleet operations will be of the order of 1.5 in 100 000 years. From the information given in Chapter 7 we may estimate the chances of explosion following the fire at 30 per cent. Thus, the chances of an explosion are one in a million years.

For such an explosion to kill an appreciable number of people, it would need to take place in a crowded location of a built-up area. The chances of the tanker being in such a place are highly conjectural, but if it were 20 per cent and as many as fifty people were killed the societal risk only becomes ten every million years.

It has previously been noted that the actual deaths in the UK arising from vehicles carrying hazardous loads approximates to two a year, and this from the operations of some 18 000 registered vehicles. Thus, the 'normal' traffic accident fatalities to be expected as a result of the operations of the fleet of ten vehicles is of the order of one every thousand years. Clearly the hazardous explosion scenario that has been postulated is many orders of magnitude less than this and represents such a low risk that it may be classed as negligible.

It may also be noted that since there is a variety of transport routes and an accident might occur at any point along them, the risk to an individual at any particular point is very much lower in relation to the societal risk than is the case for most fixed installations, although it will be recalled that the discussion of Figures 8.6 and 8.7 made a similar observation about Sizewell 'B'.

Precautionary measures

In the circumstances of such random and rare occurrences the evaluation of probabilities is not very helpful, and management must simply set out to ensure by every practicable means that appropriate measures are taken to avoid any untoward catastrophe. In the case of transport systems these would be considered under the following headings:

1 design and construction of special systems, containers or vehicles;
2 vehicle inspection and maintenance;
3 education and training of drivers;
4 prescription of routes.

The most common feature of the vehicles used for hazardous loads is a vessel designed to work under pressure, or refrigeration. The extent to which such a design has to meet the special and specific requirements of a particular substance depends primarily upon economic factors largely associated with the volume of the activity, and the scope for contract hauliers. The special ICI pipeline system carrying ethylene from Teesside to the west coast and to Scotland is an exception to the general experience which rules against the high cost of specialized systems. Nevertheless, some hazardous substances such as radioactive wastes are transported in highly specific containers which have been the subject of special design studies and impact tests to simulate collisions. Thus, in the case of chlorine containers industry endeavours to ensure that:

1 the containing steel is of an appropriate quality;
2 the welding is by an approved method by certified welders and with 100 per cent inspection using radiography and other non-destructive procedures;
3 that approved and standardized valves are fitted together with a device which prevents leakage in the event of failure of the main discharge valve;
4 test pressures are greater than are required by the regulations;
5 protection of critical equipment with sufficiently thick steel to give adequate impact resistance;
6 strict rules are followed when filling and emptying containers.

In addition to this attention to design, construction, and operation, the industry in the UK provides an emergency response to untoward incidents. This scheme, known as Chlor-aid, has already been described in Chapter 4.[7]

The design of pressure vessels has been discussed in Chapter 4 and will not be enlarged upon here. Apart from petroleum spirit, some 80 million tonnes of chemicals are moved by road every year but only half goes in tankers, the rest is taken in drums, cylinders and bottles; and for powders, crystals and pellets, in sacks and cartons. The vast majority of the packaged chemicals are not dangerous at all and certainly not likely to constitute a major hazard. Over half those chemicals are not moved in tankers.

In the UK there are many legal requirements covering every aspect of the movement of chemicals. The most well-known feature of these must be the Hazchem warning sign. Displayed on the rear and side of every tanker carrying a hazardous load, it is a requirement of the Dangerous Substances (Conveyance by road in road tankers and tank containers) Regulations 1981. It began as a voluntary practice under the auspices of the Chemical Industries Association (CIA), and an example appears in Figure 10.1.

The pictorial illustration in the diamond sign indicates to the general public what kind of chemical hazard is possible. Examples of alternative pictorial illustrations are inset. The rest of the notice gives coded information to emergency services such as the fire service and police, as well as a telephone number where further expert information can be found. Also inset in Figure 10.1 is an example of the code's interpretation so as to specify the chemical and the way it should be treated with special reference to spillage, fire-fighting, and so on. The CIA has also introduced a voluntary scheme under which similar notices are displayed on vehicles carrying low-hazard chemicals. These do not have a diamond sign, but are otherwise similar in scope.

Small containers and packages of hazardous material are similarly required to have an appropriate label, and where material is shipped across frontiers a set of accompanying documentation is also required. Such documentation involves:

1 UN classification numbers, recognized name in English, French or German;
2 maritime codes or air transport codes used to process shipments by means of checklists, and so on.

The problems of handling such containers and the associated documents at docks and international airports can be severe and there is a growing tendency to rely on computer-based systems

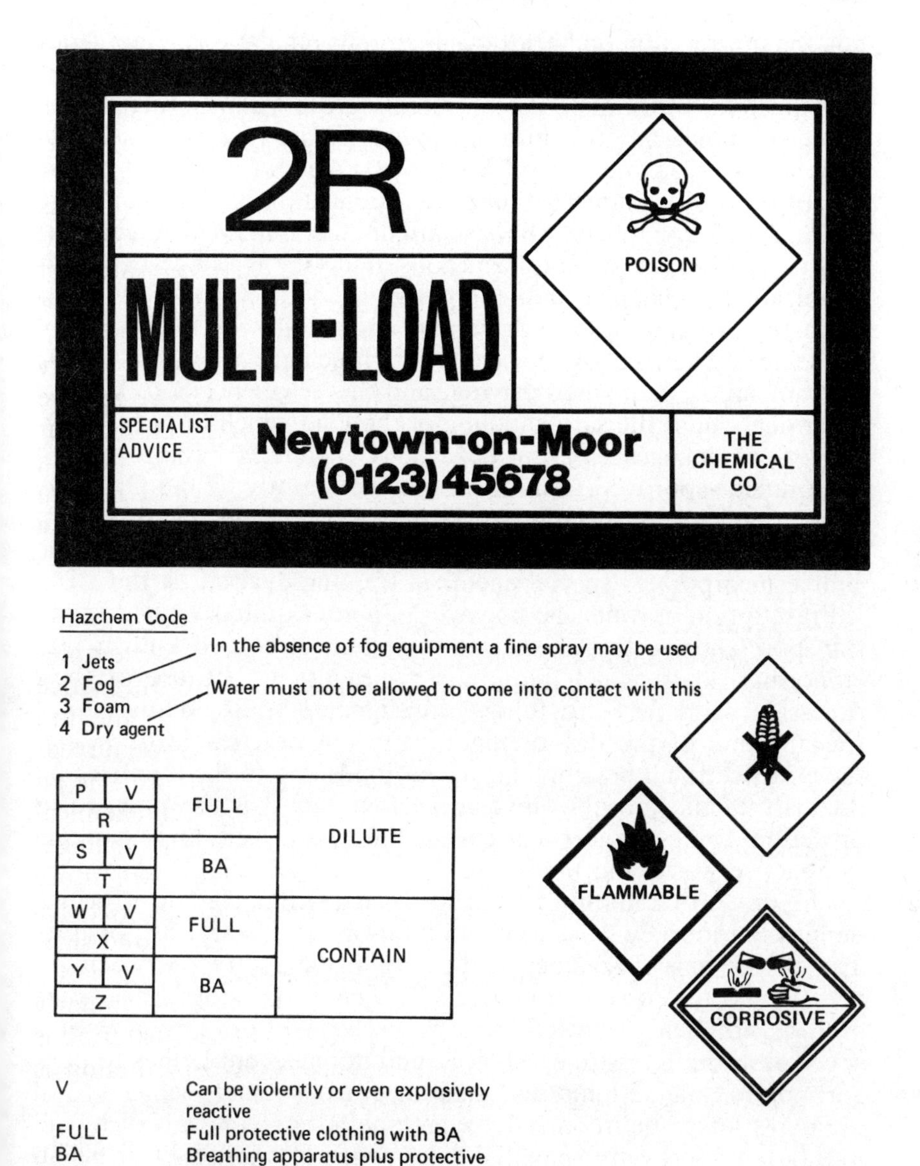

V	Can be violently or even explosively reactive
FULL	Full protective clothing with BA
BA	Breathing apparatus plus protective gloves
DILUTE	May be washed to drain with large quantities of water
CONTAIN	Prevent by any available means spillage from entering drains

Figure 10.1 *The Hazchem warning sign*

holding information on hazardous materials regulations, emergency response procedures, and medical guidance.

Inspection and maintenance systems are a vital feature of the measures which any organization must operate to avoid risks, not just in transport but throughout the range of activities which constitute major industrial hazards. Education and training are similarly widely applied. All these are discussed in the next chapter. Where road transport of hazardous material is concerned it is important to recognize that the driver is required to know much more than merely how to drive and understand the mechanics of his vehicle. The risks of an accident are much greater when loading and unloading than when driving, and the driver needs to know a great deal about the safe handling of the load which he is carrying. Drivers of vehicles carrying hazardous chemicals have to pass a government-approved training scheme run jointly by the CIA, the Road Haulage Association, and the Road Transport Industry Training Board. In addition to appropriate driver training the course incorporates special chemical training devised by the CIA.

Prescription of routes is, however, a matter quite specific to road transport and worthy of special comment. It is well known that 'abnormal' loads which require a police escort because of their excessive width have to follow a designated route, perhaps at a specific time of the day or night. It is not practicable to extend this practice to all possibly hazardous loads due to the number and diversity of the journeys involved. However, it is the practice to prescribe routes in certain cases, the most well known being radioactive waste and munitions.

The driver of a munitions truck is issued with a route card which defines the journey so as to avoid built-up areas so far as possible. The driver must also keep a distance of 75 metres between his vehicle and the one in front. He is subject to parking restrictions, such as not being allowed to park his vehicle on a road with a speed restriction less than 40 mph, and in any event he has to park with more than 50 metres' clearance from other vehicles. All accidents must be reported to an operations centre which will initiate any necessary remedial action such as transfer of the load to another vehicle.

All drivers of chemical tankers with hazardous loads carry a transport emergency card (known as a 'Tremcard') which provides instructions for dealing with an emergency. The cards are drawn up to a prescribed format; are kept in the driver's cab; and are always in the language of the country through which the vehicle is travelling. They identify the chemical and provide information on how to act if the chemical should escape from its container.

Miscellaneous information

An interesting perspective on transport risks is provided by some of the accident statistics relating to transport operations. An analysis of rail transport incidents[8] involving flammable liquids and gases in the USA showed that:

1. 632 incidents concerned liquids, 450 involved gases;
2. 6–7 per cent of all incidents resulted in some ignition;
3. 70 per cent of all incidents resulted from loose or defective connections and container fittings rather than a transport accident;
4. 20 per cent of all incidents resulted from a transport accident to the vehicle.

In the UK in 1981 official statistics show that:

1. there were 33 770 fires to road vehicles,
2. of which only 881 were due to crashes of the moving vehicle or road traffic accident;
3. of these 677 were to private cars,
4. 147 to motorcycles, and
5. 13 to goods vehicles;
6. there were 33 344 road traffic accidents to goods vehicles;
7. these resulted in 9932 casualties,
8. of which 7387 were in other vehicles,
9. 854 were in the goods vehicles, and
10. 1691 were pedestrians.

Official statistics in the US for 1977 showed that:

1. 42 per cent of road traffic accidents to 'carriers of property' were due to collisions with private cars;
2. 13 per cent were due to collisions with other trucks;
3. 10 per cent were due to a mechanical defect;
4. 9 per cent were due to a collision with a fixed object;
5. 2 per cent were due to running off the road.

The importance of care and maintenance is shown from the UK Department of Transport 1981 data relating to fires in road vehicles of which there were 33 770, but only 881 of these were due to crashes (2.6 per cent):

1 26.6 per cent were due to electric wiring faults;
2 25.1 per cent were due to oil and petrol spilling on to hot components;
3 22 per cent were due to deliberate action;
4 3.4 per cent were due to smokers' materials;
5 12 per cent were due to other identified causes;
6 8.3 per cent were not explained.

It has been demonstrated on a worldwide basis that the hazards arising from the transport of dangerous substances cause a comparable number of deaths to that from fixed installations. There does not, however, seem to be as much concern from the public about such transport deaths. Possible explanations for this state of affairs may include:

1 The location of the hazard is much more random than is the case for fixed installations. There is a relative lack of a geographic focal point for the expression of concern or protest.
2 When compared with the risk assessments for most fixed installations, the individual risk is low relative to the societal risk.
3 Road traffic, which is the major category of hazard, does not greatly alarm the public, even though it provides many casualties on its own account irrespective of hazardous loads. Perhaps a large section of the public is aware that privately owned cars are themselves a principal source of the trouble.

References

1 Kletz, T. A. and Turner, E., *Is the number of serious accidents increasing?*, ICI Safety Note 79/2B, London: Chem. Ind. Assn., 1979.
2 Kloeber, G., Cornell, M., McNamara, T. and Moccati, A., *Risk assessment of air versus other transport modes for explosives*, US Dept of Transport PB80–138480, December 1979.
3 Meslin, T. B., 'The case of chlorine transport in France', *Risk Analysis*, 1, p. 137, 1981.
4 *Reactor Safety Study*, App. VI, Section II WASH–1400, Washington, DC: US Regulatory Nuclear Commission, October 1973.

5 Anderson, P. N., *New developments in UKHIS labelling*, Teesside Polytechnic Symposium, April 1976.
6 Chemical Industries Association, *Chemicals on the move*, London: CIA, 1985.
7 Carr, B., *Chlor-aid – the intercompany collaboration for chlorine engineers*, p. 113, 1985 International Chlorine Symposium, London: Soc. Chem. Ind./Ellis Horwood, 1986.
8 Kloeber *et al.*, op. cit. note 2.

11 The mitigation of hazards

In spite of their potential for harm, it is often necessary to proceed with the construction and operation of hazardous industrial installations to provide benefits for the community as a whole. It is then necessary to set up a defence in depth against the worst possible consequences, although it must be admitted that elimination of all possible risks is an impossible goal. Such defences usually have disadvantages, however, such as inconvenience or extra costs. A view will have to be taken of their appropriateness; this view is not likely to be absolute and may change with personal circumstances and experience.

Of the accidental deaths in the UK listed in Table 1.5, a large proportion occurs at home; the greatest number from falling down stairs. Nevertheless, most people opt for houses for their families, although for elderly people the balance of advantage clearly lies in a bungalow.

Some of the considerations affecting public policy in regard to this balance and the consequential legislation have been discussed in Chapter 3. This chapter is concerned with the defences that can be made. They fall under four headings:

1 design and construction procedures;
2 maintenance operations;
3 education and training of staff;
4 emergency plans.

Each of these will now be discussed in turn.

Design and construction procedures

Prevention is always better than cure, and although it is not likely to be possible to eliminate risk altogether it is possible to go a long way towards this goal by sensible design and sound construction of process plants.

In the early years of the process industry all operations were monitored and controlled by human operators. To enable the operators to do their work effectively there was a need for the reactions and processes to proceed at a pace compatible with the operator's response, and to provide an appropriate buffer against dislocations in procedures such as would happen at times of shift change, for example. The human operator works most reliably when he is able to do things in sequence, one at a time; he is less reliable when he has to watch over and do things almost simultaneously. For these reasons there was a tendency to design and build plants based on a sequence of batch operations with large inventories both in the reactors and in buffer stores. The large residence times in the vessels provided plenty of opportunity for the human operator to take corrective action as and when necessary. Of course, with very slow reactions inside large vessels the watching attendant might become so relaxed as to fall asleep, with disastrous consequences!

Trevor Kletz[1] has provided a dramatic illustration of this former era, from the manufacture of nitroglycerine. It used to be made from glycerine derived from whale oil put in a batch reactor and acted upon by a mixture of nitric and sulphuric acids. The reactor was simply a large pot containing 1 tonne of material, fitted with a mechanical stirrer and cooling elements. The reaction is strongly exothermic and if the heat of reaction is not removed by cooling and stirring an uncontrollable oxidation is quickly followed by a violent explosive decomposition of the nitroglycerine. The operator had therefore to keep an eye on the stirring and cooling systems and watch the temperature closely. If all went well he did not have much to do until the reaction was completed, but this took two hours and he was likely to fall asleep in the meantime. However, if the reactor were to explode he would have no chance of survival – indeed the whole plant was usually destroyed in such circumstances. To prevent such a catastrophe the operator was made to sit on the now legendary one-legged stool.

The modern process employs a continuous reactor and the reaction time has been reduced from two hours to two minutes.

The acid is pumped at high speed through an injector, drawing the glycerine in through the side, the system being designed to ensure the right proportions. The turbulent jet provides good mixing and the reaction is largely self-regulating, but simple automatic controls are provided to shut the plant down automatically in case of malfunction, while the quantity of nitroglycerine is so small that the attendant operators are safely protected by a single blast wall.

The advent of automatic systems of process control throughout the chemical processing industry has greatly stimulated the development of continuous production methods, and this in turn has been accompanied by a drastic reduction in the size of inventories per unit of throughput. This lowering of the inventory in hazardous processes is a very significant factor when safety is assessed. Although the move from batch to continuous, from manual to automatic, may have seemed spectacularly rapid over the years since the end of World War 2, it should not be forgotten that automatic governor control of windmill speed was featured in the Middle Ages, incorporated by Watt into steam engines at the start of the Industrial Revolution, and that automatic temperature control of process heaters was successfully introduced in 1831.

The successful wholesale adoption of a new design concept into manufacturing industry is a complicated process in itself, and many factors influence the rate at which changes take place. Neither should it be supposed that technology itself can always provide a more satisfactory alternative to an established practice. The tragedy at Bhopal described in Chapter 2 centred upon the manufacture of methyl-isocyanate, and this is just one of a family of isocyanates which are widely used in the process industry for the manufacture of other products in everyday use. Almost all this isocyanate is made using phosgene, which in turn is derived from chlorine. These are all very nasty gases as Table 2.2 makes clear. Both chlorine and phosgene were used as weapons in World War 1. Much research has been carried out on the development of alternative processes but nothing has so far emerged which seems sure to replace the traditional methods. However, recent installations have reduced substantially the quantity of phosgene held in storage, and in one case no liquid phosgene is stored at all, the gas from the production unit flowing directly into the consuming unit.

The main driving force behind the adoption of continuous, automatically controlled unit processes has been the great financial saving made possible by the reduction of capital cost in construction and labour costs in subsequent operations, together with the possibilities for more uniform product quality. These lower costs went hand in hand with market growth and during the 1960s it seemed to all concerned that the more investment that was made, the more was the return. At this time there was a corresponding increase in unit plant size accompanied by a reduction in the unit capital cost per unit of throughput. Further economies accrued from greater process efficiency made possible by continuous operation and, of course, from the labour saving.

While the increased scale of operations did not involve any new innovation in a scientific sense, severe technological and engineering difficulties often arose during construction due merely to the increase in size. Delays in construction and commissioning often arose because of these unforeseen technological problems, and because of the large amounts of capital at risk and the cost penalties of coming late to the market, many sectors of the process industries suffered financial loss.

Throughout the 1970s the industry tended to play safe with its money, freezing its designs at a very early stage in the project programme, not merely in respect of scientific innovation but in every possible way including unit size, so that only thoroughly tested and proven systems were commissioned. This led to a marked improvement in the completion of major products to time and cost, but has placed a constraint upon the development of new designs even though they may have potential for greater safety.

The next chapter considers such cost implications further, with special reference to the assessment of the cost of safety measures. Suffice to say at this point that whereas in the 1970s most of the large chemicals manufacturers were content to press forward with substantial investment commitments on the basis of confident market forecasts rooted in the strong market growth of the 1950s and 1960s, this is no longer the case. A worldwide check in market demand has led in the 1980s to overcapacity on the supply side and uneconomic plant working in the USA and Western Europe. This has brought on large-scale redundancies of men and facilities, and it seems most likely that in the immediate future it will on balance be more advantageous to keep older plants in operation, and there will be fewer incentives to build plants incorporating the latest scientific developments.

It therefore appears that for the time being there will be greater opportunities for designers in removing the bottlenecks from existing plant systems than in creating new processes, and in making marginal improvements to existing plants by adding on new features, such as improved automatic control equipment to improve reliability and safety.

Unlike the situation described earlier in respect of the manufacture of nitroglycerine, most of the materials handled in the oil and chemicals industries are not flammable or explosive in themselves but only when mixed with oxygen (for instance in air). It may therefore have been somewhat misleading to introduce in Chapter 8 the concept of TNT equivalence since, whereas TNT can explode by itself, substances such as petrol and LNG cannot burn or explode unless they are mixed with air. The majority of petrochemical processes operate under pressure and the process materials are often stored under pressure so that air is not likely to find its way into such vessels. Hence the problem of preventing fire and explosion hazards, as with the prevention of toxic hazards, often reduces to that of simply preventing leaks.

Nowadays automatic detection, alarm and control equipment plays a leading part in the prevention of hazards. Ideally it should be fully integrated into the automatic process control system for the operation of the plant as a whole. Fully automatic process control has been a feature of the petrochemicals industry for many years, but it is a rapidly changing scene so far as the choice of equipment is concerned. There may have been as many as four complete changes in the character of an installation during the career of a typical instrument engineer. At present all the running is being made by electronic systems based on microcomputers, although the technology of the final actuators is rather more conventional and less subject to change. Of particular concern is the need for properly planned and competent maintenance of this equipment and for its way of working to be thoroughly thought through, so as to provide a meaningful and satisfactory role for the human operators and supervisors who remain in charge.

Although there have been many obvious benefits to the working conditions arising from the tremendous developments in automatic control of processes in the petrochemical industries, there have also been some doubts and criticisms.[2] Automation has removed most of the hazardous and dirty jobs; it has relieved the tedium from monitoring and recording slowly changing events, while fast-acting automatic valves help to mitigate much of the operator's stress in an emergency. Nevertheless, it has been said that the automatic control systems so often associated with large process plants isolate the operating staff from physical contact with the plant, reduce the workers to mere watchers of dials and pushers of buttons and inhibit a proper understanding of what is really going on. This induces boredom when everything is working in steady state, causes difficulties in the implementation of changes designed to give improved performance, and leaves people quite unprepared and overstretched during the crisis conditions of an emergency in a potentially hazardous plant. Education and training will not provide a remedy for these criticisms unless it is in large measure 'on the job', and in a potentially hazardous plant it is vital that the routine tasks of the operators require them to be fully conversant with the actions of the control system and the plant processes.

A designer's checklist for reviewing an existing process for the purposes of safety would include:

1. Identify all potentially hazardous materials.
2. Can any of them be substituted with safe materials?
3. If not is it possible to store less of them?
4. Can they be used in less hazardous circumstances?
5. Can their containment be improved or made more reliable?
6. Can the design of the containment system be improved to ease its maintenance?
7. Is there automatic and rapid detection of leaks?
8. Is there automatic isolation of leaks in appropriate circumstances?
9. Is there adequate provision for dispersal by open air construction or use of water curtains?
10. Is there adequate provision of flarestacks, scrubbers or catchpots to prevent the escape of dangerous gases or liquids in case of process emergencies?
11. Is the automatic control system designed to 'fail safe', and is it designed to be fully compatible with the job satisfaction of the process operators so that they may properly understand the part they will have to play in case of emergencies?
12. Finally, is the fire fighting equipment adequate?

Maintenance

The profitable operation of a modern chemicals plant much depends upon keeping the expensive capital equipment on stream for it to justify the capital charges which make up a high proportion of plant costs. This requires the maintenance activities to relate to production needs, rather than maintain engineering standards for their own sake. The three essential elements of maintenance management in a process industry are:

1 retention of working experience;
2 establishment of procedures, schedules and standards;
3 analysis and feedback to operations and design.

Where, however, safety equipment is added on to a plant and only operates in case of emergency, it will not be making any direct contribution to production and special arrangements will need to be made to ensure that it is not overlooked. It is desirable that the sensors, controllers and actuators needed for normal process control be incorporated as far as possible into whatever scheme is needed for emergency action.

The three elements will now be discussed in turn.

Retention of working experience

Classified records and comparative summary sheets are kept which show the lost production attributable to various units being out of service due to breakdowns in addition to the direct cost of maintenance. These make it possible to assess the cost of breakdowns and to show the likely benefits that may accrue from better maintenance or replacement of obsolete and troublesome equipment. Such records will ease the assessment of in-house and contract maintenance service. They also show whether performance is improving from one time period to another, and help management to settle priorities where the maintenance of equipment in routine use is concerned. For the reason already given, special consideration may need to be given to safety equipment which only operates in an emergency. There is of course a need for a storage and retrieval system for the engineering records, data sheets, drawings and spare parts used in the maintenance activity.

Establishment of procedures, schedules and standards

The first question to be decided is whether the equipment is to be serviced only after it has broken down, or as a result of inspection, or whether it should be the subject of a preventive maintenance programme planned in the light of the assessment previously outlined.

1 *Breakdown maintenance.* In some non-hazardous production processes, seasonal or market variations are so great, or the plant is so complex relative to the available skills, that it pays to run it until it breaks down, rather than take it out of service for preventive maintenance. A breakdown maintenance service will have to be provided, however, and in many instances, particularly with hazardous processes, there will be standby facilities, perhaps automatically switched into service as and when necessary. With breakdown maintenance it is vital that the approach to fault-finding and repair be logical. On complex equipment fault-finding procedures must be followed rigorously. Ascertaining the cause of the fault may require co-operation between maintenance and operating staff, as will testing to ensure that any repair is effective. Most complex chemical plants operate with both breakdown and preventive systems of maintenance in force.
2 *Planned preventive maintenance.* Schedules for inspection and preventive maintenance can be drawn up from figures obtained from the plant suppliers or from the records of working experience previously described. Planned maintenance usually involves different categories of maintenance work, depending on the expected life of the equipment and its time in service, the findings of an inspection, and so on. Records must be kept of the work actually executed and these should be analysed to provide data on standards, wear and failure rates, costs, time to repair, parts usage, and whether too much or too little maintenance is being undertaken. On complex plant installations it is usually helpful to base the analysis of maintenance procedures on concepts of reliability theory to provide assurance that there is an ongoing process of improving safety and reliability whenever new plant is installed and production schedules altered. These analyses should involve the active participation of production, engineering and safety managers.

Reliability analyses

Some conceptual understanding will be obtained from Figure 11.1, which shows the number of tests necessary to establish reliability with confidence. For example, if we wish to be 99 per cent confident that an article is 98 per cent reliable ± 1 per cent, over 1000 test runs must be carried out.

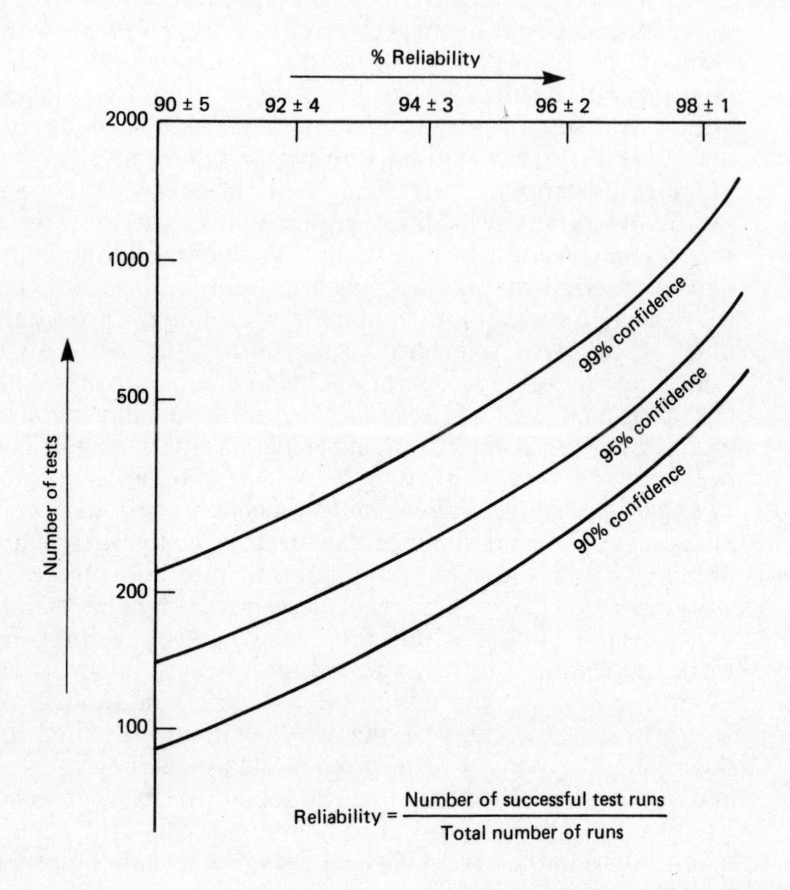

Figure 11.1 *The number of tests necessary to establish system reliability*

Figure 11.2 demonstrates the low probability of operational viability which can result from the extensive coupling of series elements which in themselves may have a high reliability. Thus, if we were to couple together fifty articles each of which is 98 per cent reliable we have a less than even chance that the complete system will perform when it is wanted. This helps to explain why in high-risk situations, novel systems take a long time to become

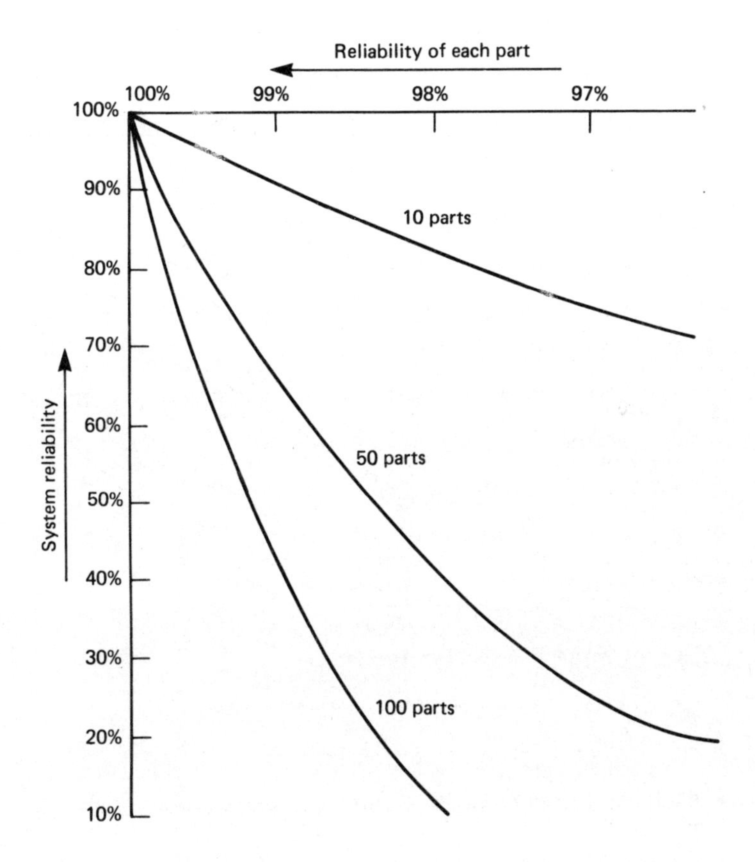

Figure 11.2 *Number of parts and system reliability*

accepted and why established and proven makes of some high-technology products, such as aeroplanes and nuclear reactors, retain their market dominance.

Figure 11.3 introduces the concept of failure rate. This is not often a true constant and averages have to be used. It defines the probability of there being no breakdown in time t provided the failure rate is constant.

Of some practical value is the concept of operating efficiency, which is dependent on the repair time as well as failure rate.

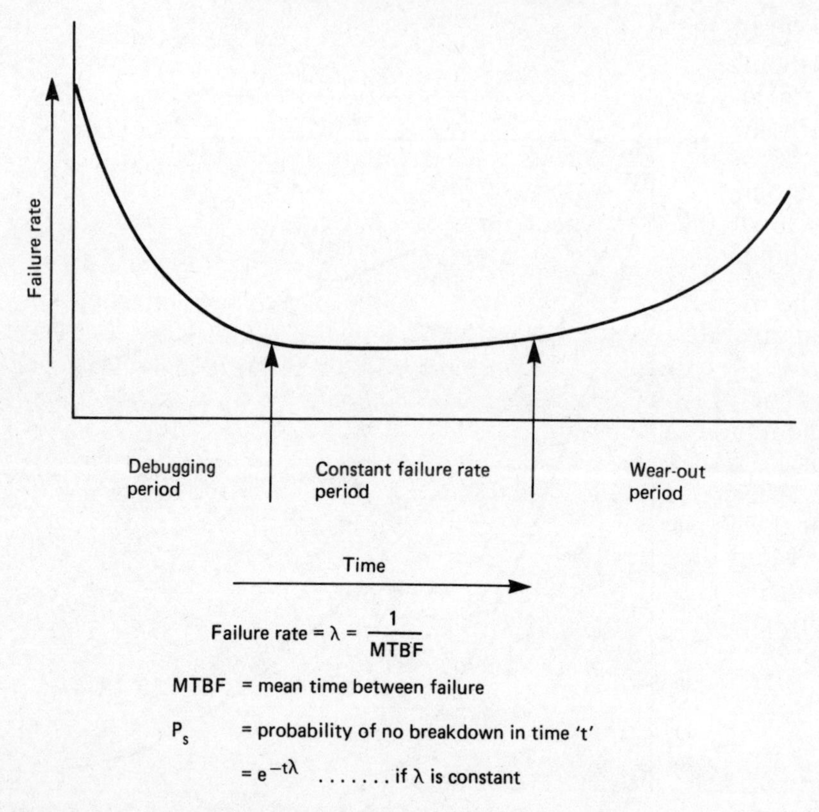

Figure 11.3 *A mortality curve: failure rate versus time*

$$\text{Operating efficiency} = \frac{\text{MTBF}}{\text{MTBF} + \text{MTTR}}$$

MTBF = mean time between failure and MTTR = mean time to repair.

Other concepts which relate reliability to production include probable availability and system hours below par in one year. For the most part, production and maintenance managements do not expect process equipment to be trouble free, and plan the disposition of their resources accordingly. The following five examples provide illustrations of the conceptual thinking.

Example 1 An electronic amplifier

An electronic amplifier is made from 100 components.

20 of the components have a failure rate ≡ 0.5 faults in 10^6 hours
50 of the components have a failure rate ≡ 0.1 faults in 10^6 hours
20 of the components have a failure rate ≡ 0.05 faults in 10^6 hours
10 of the components have a failure rate ≡ 1.0 faults in 10^6 hours

The overall failure rate of the amplifier is given by the sum of the failures in a million hours, or $20 \times 0.5 + 50 \times 0.1 + 20 \times 0.05 + 10 \times 1.0$, hence the failure rate is 26 faults in a million hours.

The probability of the amplifier having failed after 500 hours is the number of failures at that time, or $500 \times 26 \times 10^{-6}$, or 0.013, or 1.3 per cent.

Alternatively, the probability that it will be working is $1 - 0.013$, or 0.987.

More directly:

$$\begin{aligned} \text{Probability of working} &= \exp(-t\lambda) \\ &= \exp(-500 \times 26 \times 10 - 6) \\ &= \exp(-0.013) \\ &= 0.987 \end{aligned}$$

Example 2 A measuring system

A temperature-measuring system consists of three main elements – a thermocouple, an amplifier and a chart recorder. Key data is:

	MTBF	*MTTR*
Thermocouple	100 000 hours	5 hours
Amplifier	25 000 hours	50 hours
Recorder	10 000 hours	10 hours

The system efficiency is given by multiplying the individual efficiencies together:

$$\frac{100\,000}{100\,005} \times \frac{25\,000}{25\,050} \times \frac{10\,000}{10\,010} = 0.99695$$

Even if the least reliable item, the recorder, were made perfect by preventive maintenance, the availability would only increase to

0.99795, an improvement of 0.01 per cent. This is not likely to be worth bothering about and a system of breakdown maintenance would be employed. However, if there were say twenty thermocouple/amplifiers working into a common recorder on a critical process, a different view might be taken.

Example 3 Spare parts

A refinery has in use 120 diaphragm-sealed pressure transmitters. These instruments are vital for the safety and economic operation of the refinery. An average time between failures is reckoned to be two years and is usually associated with the diaphragm which would in any case be replaced since there is no point in preventive maintenance, the instruments being put into an 'as new' condition whenever a failure occurs. Replacement diaphragms take six months to be delivered, and the minimum economic order quantity is five.

In the six months' delivery time we would expect to use thirty diaphragms, and these might be the subject of a standing monthly order for five diaphragms. A reserve stock would be necessary to be sure of not running out at the end of every month. For 99 per cent availability this reserve stock is computed from the formula:

$$\text{Reserve stock} = 2.33 \times (\text{average issue rate} \times \text{delivery time})^{\frac{1}{2}}$$

which in this case is $2.33 \times (24)^{1/2}$, or 11.4, so that the average monthly stock would be:

$$\frac{\text{delivery rate}}{2} + \text{safety stock}$$

$$= 2.5 + 11.4$$

$$= 14 \text{ diaphragms}$$

Example 4 Preventive maintenance

There are six automatic batch filter units in a continuous unit process. Each filter has a miscellany of automatic valves, switches, flow regulators, and so on, which form a complete subsystem with a mean time between failure of 400 hours. Each filter is given a preventive maintenance check-out every two weeks and is overhauled completely whenever it fails. The average time out of service during repair/overhaul is five hours.

Planned inspection = 2 weeks = 336 hours

MTBF = 400 hours

Failure rate = 1/400

$$\text{Probability of work} = \exp(-t\lambda)$$
$$= \exp(-336 \times 1/400)$$
$$= 0.432$$

Probability of breakdown within 2 weeks = 0.568

Hence, average time between all overhauls for one machine is 0.568 × 400, or 227 hours.

Hence, average time for the battery of all six machines is 38 hours. The availability of the battery under full load is the same as for a single machine:

$$= \frac{227}{227 + 5}$$
$$= 97.8\%$$

If the average overhaul time were reduced to 3 hours the availability becomes 98.6 per cent.

Further similar calculations might be undertaken to compare the performance of a battery of five larger machines working with one as a standby, or even four with two in standby. (However, as Example 5 shows, the algebra becomes very tedious.) Since the cost of machines is not proportional to their capacity such an investment might be economic. Generally speaking many small machines operating in parallel would be more expensive to purchase and involve more maintenance than fewer larger machines. Whether they would give more certain production would depend on how many were in standby.

Example 5 Standby equipment

There are two ways of improving reliability and safety of equipment. The first is by designing individual items so that their failure frequency becomes so low as to be negligible, the second is by the provision of standby equipment with automatic changeover facilities coupled to an effective repair and replacement service.

Consider, for example, a simple system consisting of a pump, a pipe and a control element which provides constant pressure at the outlet. If the probability of failure of each of these three is a, b and c respectively, the reliability of the whole system is,

$$\begin{aligned} R_s &= (1 - a)(1 - b)(1 - c) \\ &= 1 - (a + b + c - ab - ac - bc + abc) \end{aligned}$$

We can improve the system reliability and safety by strict design and manufacture of the pipe to a pressure parts code so that the probability of its failure can be neglected. We can also install a standby pump and control element together with an automatic changeover system which switches the flow through the standby pump and control element if the first combination fails.

If now the probability of a pump failing is p, of a control element failing under low pressure is q, and under high pressure is r, the probability of both a pump failing and a controller failing under low pressure is pq, and under high pressure is pr.

The reliability of the system without the standby units and switch is:

$$R_s = 1 - [p + q + r - pq - pr]$$

and the probability of failure is:

$$P = p + q + r - pq - pr$$

Since the probability of the standby failing may be different we can designate the two probabilities as P_1 and P_2 for the running unit and standby unit respectively. If the probability of a switch failure when it should work is u, and of working when it should not is w, then the probability of it working correctly is:

$$P_s = 1 - u - w$$

If the time of failure of the running unit is t, the probability of both failing is:

$$\int_0^1 P_1 P_2 (1 - t) dt = P_1 P_2 / 2$$

Hence the reliability of the system is:

$$R = 1 - [(1 - u - w)P_1P_2/2 + uP_1 + wP_2]$$

Clearly the reliability of the switching system has to be a lot better than that of the pumping system to warrant the extra complexity and cost of fitting and maintaining it. Most chemical processes are fitted with standby pumping sets, but the changeover is done manually unless the safety of a hazardous process is at stake.

Figure 11.4 illustrates the kind of result one gets with a system of four parallel units in a battery. Only three are needed to obtain full production, so one is in standby. Repairs are only possible on the standby unit.

These expressions have assumed a constant failure rate, but in practice failure rate often increases with the life of the unit, although the mortality curve will be greatly affected by the maintenance regime that is enforced. The simplest way of dealing with variable failure rates is to assume constant failure rates over life increments and to calculate the reliability for each increment.

Professional institutions and manufacturers have made available a great deal of information on the reliability of process equipment. Nevertheless, the prediction of life expectancy and mortality curves in the early stages of an equipment life cycle is a very uncertain activity. This is especially so where maintenance and operation are specific to the user's environment and generic data has to be strongly supported by engineering judgement. Analysis can only set boundaries to what may be expected to happen, and these boundaries are often very inexact.

Many more aspects of inspection and maintenance to secure safety and reliability could be mentioned. However, it is hoped that enough has been said to make clear the need for all levels of management to be involved, and that the maintenance of process equipment is not the sole responsibility of a chief engineer. Safety, reliability and maintenance need to be organized on a continuous basis so that:

1 There is a vehicle for experience retention which forms part of a routine cycle for safety and reliability improvement. This should enable preventive maintenance and repairs to

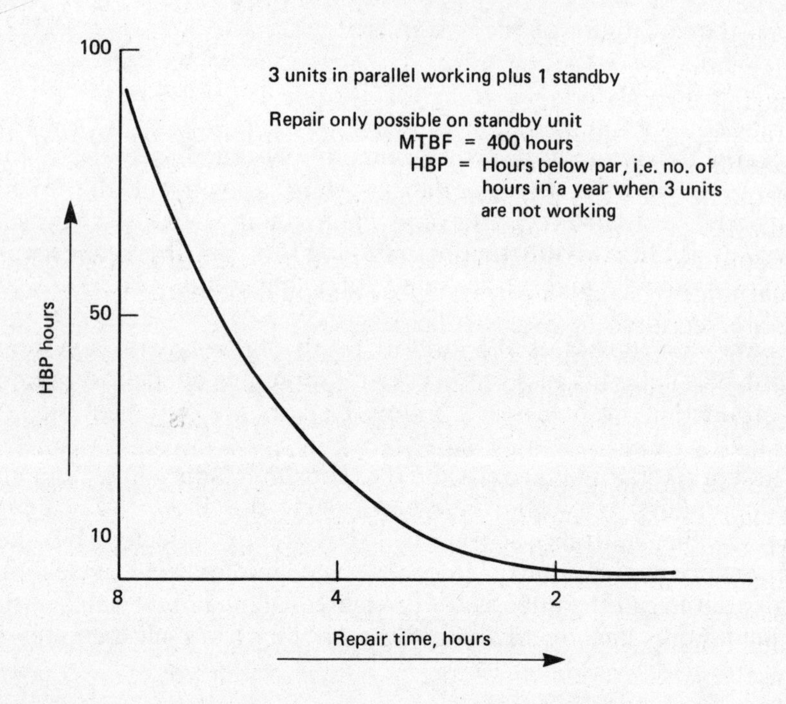

Figure 11.4 *The effect of repair time on lost production*

be assessed in measures of production performance and safety, and avoid engineering hobbyism.

2 Standards of reliability are integrated with the work of all those concerned with design, construction, process control, cost control and maintenance.

3 Accountabilities for safety and reliability are clear to all concerned so that all related activities can be properly assigned and controlled.

Education and training

The continuously evolving complexity of many petrochemical process plants has compelled industry to provide training and retraining of its staff at all levels to achieve ever more challenging productivity goals. In the course of a person's working life the technology employed, even for the same basic productive purpose, can change several times. These changes are taking place at all levels in the organization, from the slops and drains tanks to the managing director's office.

Training and education are therefore an integral part of a successful career progression, and have to be undertaken as a consequence of what a person needs to achieve in his work rather than what he can be taught. Personnel need to be screened for their aptitude and potential before they are exposed to a programme of learning, in the same way as they should be screened before they are recruited to a particular job.

Because of this emphasis on the results that have to be achieved, close involvement of an individual with his line supervisor is necessary in a process of analysing training needs. Before any training is sponsored and undertaken it is important that the individual should know why he has been selected, and what is expected in management terms as a result of the course of instruction. On completion of the training, an individual should discuss with his supervisor whether the course has met the perceived needs, how the new knowledge is to be applied and what further training may be necessary. Training establishments should themselves have procedures for the assessment of courses and lecturers by the participants, to provide a feedback of information on proficiency.

At operations and maintenance levels most companies find it necessary to establish formal training programmes keyed to such technologies as:

1 electric power production and distribution;
2 instrument maintenance;
3 machine maintenance;
4 process technology;
5 computer programming.

But a balance must always be struck between what is best undertaken by a suitably trained in-house resource and what is best purchased from outside contractors. This applies not merely to the training but to the activity itself. Formal 'classroom' training will need to be blended with 'hands-on' training. The latter may involve training simulators, workshops, laboratories and working 'on the job' alongside experienced practitioners.

In addition to the instruction courses organized by manufacturers on plant, instruments or computers, there are several short courses organized by universities and other institutions on the wider aspects of major hazard control, disposal of toxic waste and other system studies.

At the lower and middle management levels further skills will need to be developed such as report writing, statistical methods, network analyses, staff assessment and computer applications. Such work will involve participation in syndicates, working on case studies and outside visits.

Finally, at the more senior levels of management there may be a need for instruction on such subjects as cost estimation and planning, company finance, interviewing techniques, public relations, project management, information technology.

The wide scope of education and training activities in modern, technologically intensive companies provides both a challenge and an opportunity for those with a functional responsibility for safety to secure a proper integration of safety concepts into working practices. Two quotations from the writing of Lewis Carroll seem apposite.[3]

> 'Cheshire-Puss, would you tell me please which way I ought to go from here?'
> 'That depends a good deal on where you want to get to', said the Cat.
> 'I don't much care where . . .', said Alice.
> 'Then it doesn't matter which way you go', said the Cat.
> '. . . so long as I get somewhere', Alice added.
> 'Oh you're sure to do that', said the Cat, 'if you only walk long enough.'

> Alice never could quite make out afterwards, how it was they began: all she remembers is, that they were running hand in hand, and the Queen went so fast that it was all she could do to keep up with her: and still the Queen kept crying 'Faster! Faster!' but Alice felt she could not go faster, though she had no breath left to say so.
>
> However fast they went, they never seemed to pass anything. 'I wonder if all the things move along with us', thought poor puzzled Alice. And the Queen seemed to guess her thoughts, for she cried 'Faster! Don't try to talk!'
>
> Not that Alice had any idea of doing that. She felt as if she would never be able to talk again: and still the Queen cried 'Faster! Faster!' and dragged her along. 'Are we nearly there?' Alice managed to pant out at last.

'Nearly there!' the Queen repeated 'Why, we passed it ten minutes ago! Faster!'

Any newcomer to a modern operation may feel like Alice, not knowing where to begin and feeling that however hard they may try they will still be in the same state. They may even fail to recognize the goal that everyone else has worked towards. Education and training guided by participative management can do much to avoid such alienation.

Emergency plans

Planning for emergencies has to take place under two separate categories of activity:

1. by a company as a precursor to the training of on-site industrial staff;
2. by a local authority with responsibilities for the wellbeing of residents in the vicinity of a major industrial hazard.

The correct approach to the first of these is to base it upon an action plan designed to prevent the further development of an emergency on site.

On-site plans

These usually start with an identification of the various types of possible hazard and the associated warnings and alarms that need to be raised in consequence. Prospective recipients of these warnings and alarms must be given sufficient instruction and training to respond in an appropriate manner. There must be established a hierarchy of warnings and of personnel to take control at the various degrees of an emergency.

For a major emergency there will usually be a need to distinguish the persons responsible for:

1. undertaking the technical task of containing the leak and repairing faulty equipment – this may be a foreman or the chief engineer, depending on the magnitude of the problem;
2. extricating those trapped or otherwise affected, liaising with inside and outside emergency services, and generally providing support services to enable the technical team to

concentrate on its containment task;

3 management control, deciding on priorities, acting as a focal point for the dissemination of information not merely within the company's organization but also to local authorities, the police, fire service and to the media.

Emergency plans need to be thought through by all concerned and this will involve many meetings and discussions. To be effective in action they need to be rehearsed. While a great deal can be achieved with syndicate exercises, role playing and table-top games, there is no substitute for planned rehearsals on the 'live' plant itself. The co-ordination of all this preparatory work in a large undertaking may require the full-time attention of an emergency planning official and the creation of an emergency control centre.

Preventing emergencies in the process industries is therefore a subject often featured in industrial training packages, and the Institution of Chemical Engineers, for example, provides a suitable training module for this purpose complete with a video.[4]

Off-site plans

Article 7 of the Seveso Directive states that the competent authorities in the member states should ensure 'that an emergency plan is drawn up for action outside the establishment in respect of whose industrial activity notification has been given'.

The successful production of an off-site emergency plan must involve:

1 a clear understanding and commitment as to who will pay for it;
2 a clear accountability and delegation of responsibility for the planning against firm time scales;
3 availability of on-site plans with appropriate information from the manufacturer;
4 comment from the HSE on the safety case and the number of people likely to be affected;
5 co-operation between local authorities, emergency services and voluntary associations as well as with industry;
6 complete agreement amongst all concerned about the need for emergency exercises or rehearsals, and how they should be controlled and any lessons incorporated in the plans.

In the UK there has been some confusion on all these matters and consequential delay in implementation in many cases. The uneven implementation of the Seveso Directive within the member states of the EC has already been remarked upon.

Thus, in the UK while the local authorities are empowered to incur expenditure in taking action to mitigate the effects of danger to life, particularly for war emergencies, the county councils have been loathe to commit themselves to expenditure on somewhat conjectural activities covering more than a thousand industrial sites. Industry on the other hand is reluctant to bear the additional weight of costs over which it has little control and for which comparable costs are unlikely to be borne by their overseas competitors. After some discussion and delay the Secretary of State has ruled that the local authorities may charge manufacturers for their planning work (CIMAH Regulation 15), but not for the cost of emergency exercises or rehearsals.

There seems to be room for further debate arising from the conflict of opinion which is possible between those authorities who wish to plan for the worst possible accident, however unlikely, and industry which is concerned with realistic possibilities. The CIMAH Regulations leave much scope for interpretation as to what is possible, and industry may feel entitled to refuse payment for what it considers irrelevant activities. Nevertheless, in some areas excellent relationships between industry and local government have existed over many years and off-site emergency planning is of long standing. This is particularly so in those areas of Cheshire and Cleveland having large agglomerations of petrochemicals complexes.

The CIMAH Regulations also place upon industry the duty of informing the local public about a potential hazard, but local authorities may disseminate this information, with industry agreeing to pay the costs in advance. Just who should be so informed depends on the view taken of the possible hazard. The HSE takes a conservative view on matters such as safety distances, and has adopted the general distances shown in Table 11.1 for the purpose of consultation between it and the local authority. As safety cases become available more site-specific distances are likely to emerge.

There has yet to emerge any consensus as to what the public should be told. Some may feel that there should be a nationally agreed approach to this issue, others see it essentially as a local matter. A one-off distribution of leaflets, which may soon be lost or discarded, will not be sufficient and there is much to be said for constructive involvement by industry in the local community,

Table 11.1
Consultation distances

Substance	Quantity (tonnes)	Distance (metres)
Ammonia	100	250
Chlorine	50	1 000
	200	1 000
LPG	50	400
	250	600
	350	1 000
Ethylene oxide	50	1 000

with public open days regularly displaying safety arrangements and emergency plans. It certainly needs to be recognized that the minimum of information is unlikely to be enough. However, it is all too easy to over-dramatize the situation, and this must be counter-productive in the long term.

It has become clear, for example, that in the case of chlorine, as with most other toxic releases, advice to stay indoors and close all doors and windows is sufficient in practical terms. Although the industry has provided a country-wide network of emergency centres under the Chlor-aid scheme, able to respond to all sizes of chlorine spillage, and the manufacturers have well-rehearsed emergency plans, fatal accidents are virtually unknown in the UK and very few in number worldwide.[5] The most notable statistic is that fatalities from major losses of containment are with one exception restricted to inside 400 metres and more often inside 250 metres. Wholesale evacuation of people should be avoided unless very carefully thought through in advance. Moreover, provided there are appropriate medical facilities, the evidence of case histories suggests that complete recovery is possible, even from acute exposure, and that no residual respiratory disability should result.

References

1 Kletz, T. A., *Cheaper, Safer Plants*, p. 27 Rugby: I. Chem. E., 1985.

2 Withers, R. M. J., *Management aims in introducing new technology*, p. 11, IFAC Workshop, 'Productivity and Man',

Bad Boll, RKW Frankfurt, 1974.

3 Carroll, Lewis, *Alice's Adventures in Wonderland, Through the Looking-Glass*, London: Macmillan, 1865 and 1872.

4 Institution of Chemical Engineers, *Preventing Emergencies in the Process Industries*, video training package, Rugby: I. Chem. E., 1985.

5 Withers, R. M. J., *First Report of MHAP Toxicity Panel*, Proceedings of International Chlorine Symposium, London 1985, Chichester: Ellis Horwood, 1986.

12 The costs and benefits of risk prevention

The elimination of a major industrial hazard is a highly desirable aim but, as previous chapters have demonstrated, it can rarely be quantified, let alone achieved, in an absolute sense. Although intermediate steps and objectives can be precisely quantified in technical terms, consideration of the public perception can only be qualitative at the present time. For some time to come risk prevention will inevitably be a matter for a relative assessment and a judgement involving a balance of opportunities and threats, costs and benefits.

However, even the appraisal of costs in a competitive, capital-intensive environment requires a flexible and open-minded approach, so it is first necessary to discuss some aspects of standard industrial costings before undertaking a more detailed examination of the possible extra costs of risk prevention.

Standard costings

The costs of differing activities can give rise to quite different concerns, depending on their relationship to the total cost structure of an enterprise.[1] Thus in traditional manufacturing, for example, labour costs may be of prime concern; in the chlorine industry described in Chapter 3 management will be concerned about the size of its electricity bill; a polysulphone plant manager will be much more interested in his catalyst usage; a food processing plant manager in the seasonal variation in the cost of raw material; while a power plant manager will worry about the cost and calorific value of his fuel. A more quantitative illustration of these aspects is shown in Table 12.1. Care should be taken before making too close a comparison of these simplified standard costing structures.

The manufacturing companies are comparable major electric systems and components enterprises. The Japanese company is much more capital-intensive than the English concern with assets per employee about six times greater, a difference which cannot be distorted by any confusion over currency values. For each yen paid to a Japanese worker there are nearly three times as many yen in capital assets supporting him as there are pounds in the English company supporting the equivalent pound paid. But the Japanese company has had to borrow the finance to secure this end. Thus, most of the profit available for capital contribution goes in interest and service charges, even though interest rates are much lower in Japan. The English company owns 82 per cent of its assets, whereas the Japanese owns only 18 per cent; this means that much more of the profit can be distributed to the English shareholders. Over 4 per cent of sales is distributed against only 1 per cent in Japan. The Japanese company pays out much more in rent and less in taxes than its English counterpart. Whereas the

Table 12.1
Illustrative comparison of standard costings

Cost element	Manufacturing companies (%) sales				UK process industries (% sales)					
	UK		Japan		Chemical		Food processing		Electricity	
Materials	40		54		29		47		53	
Utilities	8		11		24		19		1	
Emoluments	34	*65*	23	*66*	4	*8*	7	*20*	16	*34*
Rent, taxes	8	*15*	1	*4*	20	*42*	5	*14*	4	*9*
Capital contributions	4	*8*	8	*23*	16	*35*	18	*53*	25	*55*
Sundries	1		2		1		1		1	
Net profit	4.5		1.25		6.5		3			

(Date *c.*1975 – figures in italics are percentage added value)

English company paid tax at the rate of around 51 per cent on profit the Japanese company only paid at a rate of 12.5 per cent.

These differences in attitude towards investment and labour are small in comparison to the differences seen when the three examples from the process industry are studied. The cost elements expressed as a fraction of added value for labour and capital are almost reversed as between manufacturing and process industry. In the previous chapter the importance of education and training was stressed, and in the UK electricity generating industry around one per cent of sales is reckoned to be spent in this way. It is clear that there must be severe cost pressures against such spending per employee in the manufacturing sector. Furthermore, the smaller numbers of people employed in the process activity encourage the participative style of management already noted in the process sector, and this in turn helps to secure a satisfactory return on the investment made in education and training.

There is not much significance in the variation across the line of material and utility cost elements. In the process sector these are due to wide variations in the value of by-products, the sale of which may be subtracted from the raw material costs. For utilities, the high cost of electricity in such processes as the Chlor-alkali industry has already been noted. In food processing, electricity may be generated in house at very little extra cost to the fuel already needed for steam heating. It may be a matter for debate whether this should be costed as a utility or a material element. The cost of electricity is not charged at all within the electricity generating industry. Thus, wide variations are to be expected between these lines, for which special explanations will be offered.

However, it is clear that the extra cost of risk prevention in the process industry will in the main arise from an increase in the capital costs and that an increase in capital costs of, say, 15 per cent will cause an increase on the cost as a percentage of sales of around 3 per cent. This would have a noticeable impact on profitability, and could be serious if the competition were not spending money in the same way.

Capital cost structure

To form an appreciation of the likely impact of risk control measures upon capital costs it is necessary to understand how these are made up. There is a good understanding of this subject in professional circles[2] since it is necessary for estimating and controlling the costs of capital projects. At the early feasibility stage of a large project a conceptual estimate is formulated by means of indices which convert known plant costs to planned costs for similar plant of different size and date, and by factors which convert one kind of plant cost to a different but related plant cost, or a manufactured cost to an erected cost, and so on. Most engineering departments associated with the process industries have estimating manuals which contain an appropriate range of base costs, indices and factors. These are often regarded as confidential documents. At subsequent phases of the project the conceptual estimate will be refined into more definitive estimates using information taken from manufacturing drawings, competitive quotations, and so on.

Table 12.2 provides a diagrammatic breakdown of the structure of the costs of a typically large construction project in the process

Table 12.2
A capital cost estimate

	Manufactured costs		Erection costs
Main plant items	100		50
Concrete	7.8		9.4
Structural steel	18.4		11.0
Piping	20		16.0
Electrical	12		9.6
Instruments	7		2.8
Lagging and paint	3		12.0
Bare capital costs (no civils)	160.4		
Total erection cost			110.8
Electrical and mechanical installation			101.4
Roads, yard, site preparation and management			59.6
Civil engineering contract			69
Management, engineering and commercial costs	46.6		
Plant supply contract	207.0		
Total cost		377.4	
Cost of main plant items = 100			

industry; it is based upon information obtained from a number of estimating manuals, and is therefore non-specific.

There will of course be substantial variation around these averaged costs depending upon local circumstances. The civil costs, for example, will be very dependent upon the local site, which must always be surveyed before any contractual commitment is made. A non-hazardous plant in a developing country may wish to maximize its labour costs and install a minimal amount of instrumentation. However, it may have to make extra provision for an in-house electricity generation system.

Table 12.2 shows how increases in the cost and weight of main plant items, brought about by safety requirements for standby equipment or stricter codes, will be reflected in further increased costs in many related items. Increases in instrumentation costs brought about by safety requirements for extra alarms, and by fail-safe features in control systems would not be so likely to induce significant cost penalties elsewhere.

Obviously the cost of providing extra safety will vary considerably depending on the nature of the hazardous processes involved, the completeness of the desired protection, and the methodology employed. Inspection of Table 12.2 suggests that the percentage increase on the capital cost could vary from a few per cent where only extra instrumentation is involved, to over 50 per cent where appreciable elements of the main plant items are affected. Hagon[3] has provided some illustrative data from the Mond Division of ICI; it is reproduced in Table 12.3.

The average of these costs is nearly 15 per cent, but the additional operating costs have also to be considered. A more detailed breakdown of the standard costs indicated in Table 12.1 provides an approximate ratio of 7:1 between capital costs and maintenance costs inclusive of materials and a proportion of overheads. If this is halved to make some allowance for what may be extraordinary maintenance requirements, we obtain a total averaged figure of 20 per cent on capital costs or an extra 4 per cent on sales.

Table 12.1 suggested that salaries and wages comprise around 4 per cent of sales in that particular plant of the chemical process industry. The average for the whole of the UK chemical sector at that time was nearly four times this figure – at 32 per cent of added value. It was then only 22 per cent of added value in Japan and has since fallen appreciably in the UK. Based upon an approximate cost of safety being about 50 per cent on the salaries, and taking a rough average gross emolument of £10 000 per year

Table 12.3
Costs of safety and environmental protection

Project	% of total project cost	
	For safety	For the environment
Flammable storage	50.0	—
Toxic storage/loading	7.7	11.4
Toxic gas drying	26.3	—
Caustic solids handling	3.7	7.7
Inorganic salt plant	1.5	0.4
Bleach preparation	12.5	17.2
Toxic gas liquefaction	16.8	2.4
Ester preparation	10.0	3.0
Flammable gas compression	7.1	—
Conversion to coal firing	11.8	23.5

an average extra expenditure on safety of around £5000 per year per employee is indicated.

Benefits to employees

An average employee might reasonably ask whether the benefits are worth such an expenditure as he might prefer to have the money directly for himself. This provides a starting point for an examination of benefits and an assessment of the cost/benefit relationship.

It was made clear in Chapter 1 that risk assessments start by considering the risk of early death, and that such risk from a major industrial hazard is very much greater to those employed on site than to the general public. Indeed in the UK, so far as is known, no member of the general public has ever been killed by an accident inside a chemical factory. There have been some fatalities from related transport and distribution activities, and a warehouse fire in Gateshead in 1856 killed over fifty spectators and firemen. There have also been deaths to the general public caused by accidents in closely related activities, in particular munitions work. The Silvertown explosion mentioned in Chapter 2 is such a case. The Flixborough accident (also described in Chapter 2) was a very near miss so far as the general public was concerned, and it must never be forgotten that one of the worst ever instant death tolls to the general public in the UK was directly attributable to industry. This was in 1966 at Aberfan, where 118 children and 26 adults were killed as a result of coalmining activity – normally considered to be completely isolated and safe so far as the public is concerned. Nevertheless, it is appropriate to consider first the risks of early death to employees.

Fatal accident frequency rate comparisons

The authorities in the UK, as elsewhere, provide numerous statistics on accidents to employees in the UK, and a brief example appeared in Table 1.1. A widely used statistic is the deaths occurring to 1000 workers in a given occupation over the period of their working life. This period is taken to be 100 000 hours, or a total 100 million hours for the 1000 men. It is known as the 'FAFR' (fatal accident frequency rate).

One hundred million (10^8) working hours is *c*.50 000 (0.5×10^5) working years, so that the FAFR is readily compared with figures for individual risk and the 'acceptable' figure of one in 100 000 mentioned in Chapter 1. It is sometimes suggested that loss of expectation of life should be used for comparing risks instead of the probability of death, to which it is related. It is felt that this might be easier to evaluate in money terms than death, and so make possible direct comparisons with costs. Table 12.4 gives some comparative data on these concepts for a number of industries and everyday activities which may result in instant death.[4] It may be of interest to compare these with some other death rates, shown in the lower half of the table, where people voluntarily run risks for money or for pleasure.

Some industrial risks do not cause death instantaneously; this is particularly so in coalmining and the textiles and nuclear industries, where delayed deaths may arise through the diseases of pneumoconiosis, byssinosis and cancer respectively. Neither, for that matter, does the risk from smoking. But criticism of industry, particularly the nuclear industry, is often focused on this aspect since delayed deaths are often perceived as a fate worse than instant death.

Delayed deaths

In the nuclear industry the legal maximum dose for workers laid down by the International Commission on Radiological Protection

Table 12.4

Fatal accident frequency rate and loss of expectation of life

Activity	Deaths per 10^5 persons per year	FAFR	Loss of expectation of life (in years) if starting at 20 years old: Those dying from hazard	Average of all exposed
Road traffic accidents (men)	22	11	27	0.3
Coalmining	21	10	18	2.4
Chemicals industry	8	4	14	0.1
All manufacturing	4	2	14	0.55
Motor manufacturing	1.5	0.75	—	0.02
Making clothes	0.5	0.25	—	—
Smoking 20 cigarettes a day		240	15	5.45
Rock climbing		4000		
Motorcycle racing		3500		
Skiing		130		
Offshore oil and gas working		165		
Deep-sea fishermen (UK 1959–68)		280		
Complications of pregnancy			47.5	0.01

is 5 rems per year which is estimated to give a FAFR of 25. This may seem high when compared with what is actually achieved in the chemical process industry, but it should be remembered that this is a maximum figure, and that in the CEGB there would be a local inquiry if the dose to any one individual exceeded 1.5 rems per year. In practice the FAFR is much below the average for the rest of industry. Nevertheless, in Table 12.5 the hypothetical delayed deaths for a nuclear workforce continuously exposed to the 5 rems a year are compared with the acute deaths actually recorded in the chemicals and coalmining industries and with smoking twenty cigarettes a day.

Kletz has reviewed the relative importance of short- and long-term consequences of industrial hazards and concluded that the acute and chronic risks are within an order of magnitude of each other, and that acute accidents are the more important.[5]

Table 12.5
Delayed deaths

Activity	Loss of expectation of life (days) for age (years) at beginning of exposure 20	30	40	50	60
Chemical industry (instant)	41	27	16	8	2
Nuclear industry (cancer)	68	32	12	3	0.5
Coalminers (pneumoconiosis)	876	412	154	39	6
Smoking 20 cigarettes (cancer)	1989	936	351	88	15

Thus, for a young coalminer aged twenty to twenty-four the total probability of death due to an accident is about three times that for a young man in an average occupation (that is one in a thousand per year). The risk of pneumoconiosis also increases the chance of death by a factor of 3 but will not take effect for about forty years by which time his total probability of death from all causes has risen by a factor of about 20.

Looked upon as an effect upon a potentially productive resource, the risk from pneumoconiosis does not seem as bad as the risk from a mining accident, but an individual may look forward to his retirement from a filthy occupation, and death at that time may seem to be worse than an earlier death. Probability of death is clearly not the only factor to be assessed when comparing the long- and short-term consequences.

Nevertheless, the UK national statistics reveal that of the 925 deaths per year during the period 1974–78, some 800 were due to pneumoconiosis, asbestosis, and byssinosis. These 'prescribed' diseases are restricted to a few occupations, and it seems that for industry as a whole acute deaths – at 694 per year – are the major problem. Where cancer is concerned, the Royal Society's study group suggests that only one per cent of male deaths from cancer (0.3 per cent of all male deaths) might have an occupational cause, and will include some of those due to the 'prescribed' diseases.[6]

The cost of saving a life

In the previous section the national UK statistics were quoted to show that during a lifetime's employment in an average chemicals factory there could be four deaths due to industrial accidents among a thousand workers. This comparatively low figure would certainly be much higher if it were not for the expenditure on safety measures, which was previously estimated at about 50 per cent of salaries, wages and related expenditure. If in the absence of such expenditure the number of deaths were to increase tenfold, the expenditure related to the forty lives works out as a figure of approximately £5 million per life. This is a very high figure and although it has to be treated with caution as a very rough average approximation, it does not seem in any way comparable with the costs of saving a life as practised by doctors or even road engineers (estimated at tens of thousands of pounds and hundreds of thousands of pounds respectively). An explanation of this apparent anomaly in resource allocation can be found in two quite unrelated aspects.

The first aspect has something to do with general perceptions and expectations. For the most part doctors are concerned with mitigating the effects of 'natural' disorders, and on the whole when death comes, it is singly to individuals. Road traffic accidents arise from an activity undertaken for obvious and immediate benefits, they do not often involve high multiple death tolls and the consequences are immediate and familiar. In contrast the chemicals industry is often felt to be more liable to accidents which result in multiple deaths from delayed and relatively unfamiliar and 'unnatural' effects. The management of the chemicals industry may therefore be under greater pressure to allocate expenditure for safety measures than are the authorities concerned with public health and road safety, and of course there is a much smaller constituency to look after.

The second aspect has to do with the commercial value of plant, and the cost of insuring it, in a capital-intensive industry. A major accident in a chemicals factory can be very expensive. The Flixborough explosion is estimated to have caused a property loss of over $140 million, the explosion at Pernis over $86 million. The financial cost of the loss of opportunity to manufacture is likely to be greater still in such cases. Premium payments for insurance against possible property losses are calculated on the basis of the estimated maximum loss given reasonable management controls and maintenance.[7] The insurers employ qualified engineers who carry out site surveys and make recommendations in respect of the estimated maximum loss and the premium rates. It usually pays the manufacturer to install equipment and systems in accordance with the advice of the insurers' engineers to minimize the costs of premiums. In a capital-intensive industry these extra capital and operating costs will be further inflated when expressed as a percentage of the salaries and wages; they are more appropriately assessed in terms of the capital costs.

In connection with insurance costs, it is appropriate to acknowledge that insurers consider workers in the process industries (including UK nuclear installations) to be reasonably safe. They accept such employees at normal rates for life insurance policies and contracts. This is not the case with other occupations – for example, significantly higher than average premiums are required for workers on North Sea oil rigs. These variations in life assurance premiums result from the known variations in accident rates in various occupations. It must be said, however, that insurers exclude risks from nuclear radiation from standard householders' policies in the UK.

Benefits to society

The point has already been made that so far as the process (including nuclear) industries are concerned, no member of the public has been killed by an incident in a UK factory. It is therefore impossible to relate any theoretically devised criterion of performance involving deaths to actual experience; there is no firm basis for comparison with other countries and no easy way of stipulating and assessing improvements.

Methodologies were outlined in Chapter 1 for describing hypothetical societal risks in terms of limit curves such as the Groningen lines. These graphs of possible numbers of deaths plotted against average occurrence frequency can also be compared with what has actually occurred from a variety of other natural and man-made disasters. They invariably demonstrate that the risks from process industry are small in comparison. Opinions differ about the merit of integrating the totals beneath such curves since it is widely felt that one person dying every year for 100 years is more acceptable than 100 people once in a 100 years. This view also implies that people killed one at a time matter less. Whatever view may be taken there seems to be an even less sure basis for cost/benefit assessments relating to societal risk than is the case with individual risk.

The purpose of a cost/benefit analysis can only be to justify improvements in resource allocation to defined activities. At the present time there seems to be no relation at all between the relative risks or the costs of remedial measures relating to such national phenomena of road accidents, health risks from smoking, medical care and industrial safety. It seems unlikely that more ambitious cost/benefit assessments will greatly affect these matters.

On the other hand the Groningen concept (if not the actual lines as positioned) does provide a way of assessing hypothetical risks which is logical and in conformity with common sense. Above the upper line there lies a region where the risks are not to be accepted. Below the bottom line there is a region where the risks are so trivial or so incredible that no money should be spent on reducing them. In between there lies a region where discussion is necessary between all those who have a role to play, and where measures should be put in hand to reduce and mitigate the risks as far as is reasonably practicable.

Industrial management and regulatory authorities have not only to reconcile the needs and safety of employees and the commercial requirements of the factory owners. In a democratic society they have also to balance the needs of the company and its workforce against the aspirations and perceptions of the local community. In this context industrial management will need to provide information on a variety of topics related to safety. In the longer term, however, good relationships, mutual respect and confidence are less likely to be the product of written balance sheets showing cost/benefit assessments than satisfactory personal contacts and joint working relationships built up over a reasonable period of time.

References

1 Davies, D. and McCarthy, C., *Introduction to Technological Economics*, Chichester: John Wiley, 1967.
2 Institution of Chemical Engineers, *A Guide to Capital Cost Estimating*, Rugby: I. Chem. E., 1969.
3 Hagon, D. O., *An industrial viewpoint*, Major Hazard Installations, p. 81, Seminar Proceedings, Department of Chemical Engineering, Loughborough: LUT, 1984.
4 Reissland, R. and Harris, D., *New Scientist*, p. 809, September 1979; The Royal Society, *Risk Assessment*, Study Group Report, p. 84, 1983.

5 Kletz, T. A., 'Now or later? A numerical comparison of short- and long-term hazards', *I. Chem. E. Symposium Series 80*, 1, A1, 1983.
6 The Royal Society, *Long-term Toxic Effects*, Study Group Report, 1978.
7 Redmond, T., 'Hazard assessment for insurance purposes', *The Chemical Engineer*, 406, p. 17, August 1984.

13 Conclusion

This book has sought to establish that a hazard can be described meaningfully only in terms of a risk assessment which involves a quantification of a likelihood and a magnitude. Both these quantifications will be the subject of inherent variation and uncertainty from one event to another, and assessments made on the basis of average findings need to make the limits of uncertainty clear. The general public is unlikely to perceive either of these in a linear relationship to risk.

Chemical process hazards inevitably span a whole range of magnitudes and likelihoods, but the highest likelihoods may have a trivial magnitude and the largest magnitudes an incredibly low likelihood. The ultimate impact upon a target population will be the result of a train of events each of which will have its specific cause and effect likelihood and magnitude relationship, and each will be subject to its own specific uncertainty. If a systematically optimistic or pessimistic view is taken of each event in the train, the accumulated estimate may be so optimistic or pessimistic as to be absurd. A realistic hypothesis should therefore be made of each event, and any desired degree of optimism or pessimism should be introduced subsequently in the light of the various uncertainties and confidence levels.

There is less uncertainty when estimates are made of the more serious effects upon large numbers of people than when estimates are made of the extent of hazards which have only a marginal effect upon a few people. This is exemplified by the toxic effect of a slowly dispersing cloud of poisonous gas, where there are inherent variabilities in the extent of the dispersion to low concentrations and in predicting the consequences when only a few people are likely to be affected.

Public authorities must be concerned about the welfare of everyone, including those marginally likely to be affected. In the light of the uncertainties they are bound to take a conservative view; but because of the inherently low levels of confidence their recommendations on safety distances, for example, may be thought unduly pessimistic or even alarmist when set beside a realistic estimate. Nevertheless, the general public has a right to expect a public authority to take a cautious view. It should be noted, however, that public opinion is itself variable and subject to change. There is some evidence from attitude surveys that people who live close to a potential hazard are less concerned, perhaps because they are better informed or because they are more conscious of benefits, than a nationally representative sample living further away. But there is also a historic trend in people's attitudes showing an improved expectation for life generally; this has been brought about largely by the improved control over the living environment made possible by industrial technology.

Some early assessments made of the risks from industrial hazards postulated very large numbers of people at risk because of the wide levels of uncertainty; levels which inevitably led to disagreements and debates between experts. Since then the increased research and investigative activity stimulated by the professional institutions and by the establishment of official bodies such as the Major Hazards Advisory Unit of the UK Health and Safety Executive has increased understanding and improved the scope for consensus. Estimates for the frequency of pressure vessel failure show a converging trend and the range of hazard distances given by the more recent mathematical gas dispersion models has shortened and narrowed appreciably. There has been an agreed and improved understanding on the lethal toxicity levels of some industrial gases, notably chlorine and ammonia. All this in turn has permitted a less conservative view to be taken and more realistic advice to be given. For example, in recent escapes of hazardous gases advice to local people to stay at home and close all doors and windows has been followed, rather than earlier precepts of wholesale exodus.

There are of course a number of areas where further research and investigation will provide a greater understanding and opportunities for improved performance. Estimating the size and frequency of the hazardous release seems set to remain a principal source of difficulty. Site surveys will inevitably involve many man-hours if the work is to be thorough, and aggregation of the information obtained will be a matter for judgement although, as mentioned in Chapter 5, the methodology of 'expert' computer-programmed systems is now being developed to ease matters. A number of mathematical computer-based models have been developed to ease dispersion calculations and these have been checked against the findings of large-scale field trials. There is a growing consensus amongst predictions of average results and this has been utilized in the scaling law suggested in Chapter 6. However, considerable doubt remains about the allowance to be made for weather conditions, and there is no immediate prospect of predicting dynamic consequences of turbulence. Thus, it is not possible to predict the outcome of a specific release with certainty, even though the average result of a hundred such releases may be agreed. Much is known about damage relationships, particularly where radiation and blast are concerned. However, the damage and injuries caused by fire are extremely variable and averages based on historic information is all that can be attempted at present. Toxicity relationships based upon animal experiments have been determined for most of the common industrial gases, but there is still much uncertainty when these are extrapolated to low concentrations and small numbers of human beings. In spite of all these weaknesses, however, the methodology of risk assessment is now able to provide valid comparative assessments over a wide range of circumstances. Nevertheless, there remains a gap between the theoretical understanding and the practical measures adopted to improve the safety of industrial operations on the one hand, and the kind of information released to the public on the other. There is here an ongoing task for industrial management to keep a watching brief over the developing methodology and the prospects for improvement. The position of management has often been stressed in this book, not least in its interlinking role.

The highly competitive worldwide market and the lead which technology can provide have not encouraged openness in the management of the chemical process industry. Nevertheless, the professional institutions and the manufacturers have over many years provided opportunities for information exchange on key topics, often on an international scale. Matters related to safety have figured prominently in this regard. The promotion of the

Hazchem Warning Code by the Chemical Industries Association among its members in the UK and its eventual official adoption throughout the EC is a good example of such work; there are many others. The disciplines which such bodies often seek to impose on the behaviour of their membership may be seen in strong contrast to the lack of any corresponding efforts by some pressure groups and media units. Throughout the EC reactions from industry to the new directives have been mixed. In some countries industry is very reluctant to disclose any quantitative information on the grounds that over-cautious and conservative advice from public authorities will only enhance public fears rather than allay them, and that a lack of tangible returns for what is seen as questionable expenditure in a competitive situation is more likely to be disadvantageous than beneficial to the local community in the long term. In such industrial circumstances whatever expenditure is necessary to satisfy the insurers may be thought sufficient.

With the increased knowledge and awareness on all sides has come a growth in the formulation of statutory regulations and the powers of public authorities throughout the EC. Indeed, the widespread agreement upon conservation and safety directives amongst member states of the EC is widely regarded as one of its more useful and successful accomplishments. While the practical implementation of these directives among the member states is very uneven, there can be little doubt that they represent a movement in the right direction, and in the long term must prove beneficial to all in the community. The twin policy concepts of separation of hazard source from hazard target and regulation and inspection of factory activities are well understood and long established in the UK, while the 'best practicable means' approach is well suited to the implementation of measures which often only follow from a possibly incomplete and somewhat indeterminate assessment of expected risks and benefits. While the rigid application of absolute standards will be appropriate in many instances, in others it will not.

The formulation and rehearsal of emergency plans have long been the practice in many UK companies; they are now a mandatory requirement of the EC directives, not merely within the confines of factory management but also for the local authorities. The latter's experience in these matters is very variable as is the contribution they may be able to make. In the event of a major emergency involving the local community, being prepared to respond to the need for co-ordinated activity is vital, and in some districts responsibility relationships have already been tested in large-scale rehearsals involving co-operation between industry and the public services. There is, however, a continuing debate about the character of advance information given to the public, and there is some concern over the activities and social responsibility of less well-known operators in the fields of transport and warehousing.

Cost/benefit assessments cannot be taken too far at the present time, although cost information will always be required to keep matters in a proper perspective. Many factors need to be taken into account when assessing industrial risks, but it should not often be necessary to hold such a major inquiry as has occurred over the Sizewell 'B' PWR. If it were to be regarded as a precedent for securing further public satisfaction, a much improved methodology and terms of reference for such inquiries need to be established.

It seems inevitable that there will remain many grey areas and unquantifiable issues between public perceptions and expectations and whatever measures industrial management may feel justified in undertaking as a result of any risk assessment. Both sides must recognize that there is a conflict of interest and all that can be expected is that a balance will be struck. Any proposed balance is more likely to prove acceptable to the public if it has confidence in the integrity and competence of all levels of the company's management. Such confidence seems more likely to be won as a result of the enlightened involvement of industrial management in local affairs than by any other form of public relations activity. Public fears about some aspects of environmental pollution, particularly nuclear hazards, seem set to remain a national and international matter, however, and purely local efforts, however successful they may appear to be, are not likely to provide a complete answer.

Index

Aberfan disaster 230
acceptable risk 4–6
Advisory Committee on Major Hazards xi
  reports of 51–2
American Society of Mechanical Engineers (ASME) Code 70–2, 96–7
  management under 72–7
ammonia
  dispersion of 121–2, 124
  frequency/fatality magnitude data for 17, 94, 95
  inhalation of 6–11
  in Potchefstroom disaster 28, 122
  safety distance for 222
  storage of 79–80
  toxicity of 5–6, 29, 31, 41, 62, 160, 164, 165, 238
    animal studies 10–11, 162
  transportation of 27–8
animal testing 10–11, 49–50, 161–3
  species differences in toxicity 162–3
API (USA), provisions of standards and recommended practices by 69
asbestosis, deaths due to 20
Atomic Energy Agency 77
Atomic Energy Authority 12, 82
  Safety and Reliability Directorate of xi, 96
Austrian wine adulteration incident 44

Bhopal disaster xii, 3, 20, 30–4, 41, 54, 159, 202
  causes of 33
Boiler and Pressure Vessel Committee, administration of ASME Code by 72–7
bromine, toxicity of 31, 161
butane, dispersion of 123
byssinosis, deaths due to 20

Canvey Island Reports xi
  hazard dispersion estimates in 112, 113, 118, 119
    hydrogen fluoride 123
  population data in 175, 180, 181
  process plant information in 56
  risk assessment in 17, 30, 61–2, 83, 128, 158, 184, 185, 186
    ammonia release 29, 169–70
CEFIC (Chemical Companies' European Federation) 54
Checotah highway explosion 43
Chemical Industries Association 69–70, 191, 194, 240

training scheme run by 196
Chemsafe scheme 70
Chernobyl disaster 3, 11, 12, 36–41, 51, 157
causes of 37–9
Chlor-aid scheme 70, 194, 222
chlorine
dispersion of 122–3, 124
frequency/fatality magnitude data for 94
incidents involving 29–30, 70
manufacture of 49
storage of 101
toxicity of 13, 28–9, 31, 41, 160, 164, 165, 238
animal studies 162
transportation of 44, 193–4
uses of 48, 202
Chlorine Institute (USA) 69–70
classification of major industrial risks 23–47
explosions 24
fires 24–7
Feyzin disaster 25–6
Flixborough disaster 25
San Juan Ixhuatepec disaster 26–7
nuclear hazards 36–41
Chernobyl disaster 36–41
Three Mile Island incident 36
on and off-site hazards 41–4, 128–30
Checotah highway explosion 43
Spanish campsite disaster 42–3
product hazards and waste disposal problems 44–5
toxic gas hazards 27–36
Bhopal disaster 30–4
Montanas runaway train disaster 29–30
Potchefstroom disaster 28–9, 122
Seveso disaster 34–6
see also explosions, fires, nuclear radiation, toxic gas hazards
CONCAWE (Oil Companies' European Organization of Environmental and Health Protection) 54
Control of Industrial Major Accident Hazard (CIMAH) Regulations 54–8, 221
Control of Pollution Act (1974) 54
Cremer and Warner risk assessment studies 92–5

damage relationships 138–67
damage from explosions 139–51
classification of blast casualties 141–3
primary blast relationship 143–6
secondary blast relationship 146–50
TNT-equivalent concept 104–5, 150–1
damage from fire and radiant heat 151–4
nuclear radiation damage154–9
major nuclear hazards 157–9
toxic gas hazards 159–65
effect of shelter 165
inhalation rates 164
toxic mechanisms 160–1
uncertainty of experimental data 161–3
vulnerable people 164
Dangerous Substances (Conveyance by road in road tankers and tank containers) Regulations (1981) 194
Davenport list of vapour cloud explosions 86–7, 88, 89, 128, 129
DOW Chemical Company (USA), provision of recommended practices by 69

ethylene, pipeline for transport of 193
ethylene oxide, safety distance for 222
event trees 82, 171–3
explosions 24
chances of 127–37, 192
fatalities 131–7
hydrocarbon release 131–7
off-site 128–30
on-site 128, 129
damage from 139–51
classification of blast casualties 141–3
primary blast relationship 143–6
secondary blast relationship 146–50

TNT-equivalent concept 104–5, 150–1
see also Flixborough disaster, Silvertown explosion
Explosives Act (1875) 54

fatal accident frequency rate (FAFR) 230, 231
fault tree analysis 82, 93
Fawcett list of accidental explosions 89, 91–2
Feyzin disaster 25–6
fires 24–7
chances of 127–37, 192
fatalities 131–7
hydrocarbon release 131–7
off-site 128–30
on-site 128, 129
damage from 151–4
Flixborough diaster xi, 25, 51, 104, 112, 229–30, 233
fossil fuels, energy production by 50–1
Freon, dispersion of 116
frequency/fatality magnitude data 2, 5, 15, 16–17, 57, 61–2, 183, 185, 234, 235
Davenport 88, 89
Fawcett 92
Kletz 90, 91

Gateshead warehouse fire 229
Groningen lines *see* frequency/fatality magnitude data

hazan (hazard analysis) 82–4
hazard assessment 168–87
event trees 171–3
methodologies of risk estimates 168–70
population composition 180–2
population density 175–80
presentation of 183–6
rapid methods 173–4
hazard dispersion, quantification of 99–126
ammonia 121–2, 124
butane 123
chlorine 122–3, 124
hydrogen fluoride 123
LNG 119, 121
propane 121
concept of equivalent mass 100–5
dense gas dispersion models 108–10, 111, 115
box models 109–10, 173
conservation models 109, 110, 171–3
downwind scaling law 99–100, 110, 112–18
neutral density modelling 105–8
values for factor a 118–19, 120
hazard release, quantification of 79–98
Cremer and Warner studies 92–5
Davenport list 86–7, 88, 89, 128, 129
Fawcett list 89, 91–2
Kletz list 87, 89, 90, 91, 188–9, 231–2
compensation for under-reporting 84–6, 127, 128
hazop and hazan 80–4
instantaneous 100–1
non-instantaneous 101–5
top event frequency estimation 95–7
hazards, mitigation of 200–23
design and construction procedures 201–5
checklist 205
education and training 216–9
emergency plans 219–22
off-site plans 220–2
on-site plans 219–20
maintenance 205–16
establishment of procedures, schedules and standards 206–7
reliability analyses 207–16
retention of working experience 206
hazard warning structure 20–1
Hazchem Warning Code 194–6, 240
hazop (hazard and operability study) 80–4
Health and Safety at Work Act (1974) 45, 59
Health and Safety Commission 51
Health and Safety Executive xi, 13, 51, 52–3, 54, 57–8, 68, 238
role of in Public Inquiries 58–9
Pheasant Wood 60–1

Sizewell 'B' 59–60
safety distances defined by 221, 222
hydrogen cyanide
toxicity of 31, 161
animal studies 162
hydrogen fluoride
dispersion of 123
toxicity of 31, 161
hydrogen sulphide, toxicity of 31, 161

Inspection Validation Centre 76
International Atomic Energy Agency 39–40, 157
International Commission on Radiological Protection 12, 40, 155, 230
Iranian food poisoning disaster xii
Italian wine adulteration incident xii, 44

Kletz list of fires and explosions (1970–81) 87, 89, 90, 91, 188–9, 231–2

LNG, dispersion of 116, 119, 121
LPG, dispersion of 222

Major Hazards Assessment Unit 53, 62, 238
methylene chloride, toxicity of 182
methyl isocyanate
application of CIMAH Regulations to manufacture of 54
in Bhopal disaster 30–4
toxicity of 31, 161
Montanas runaway train disaster 29–30
Mossmoran Inquiry 21

National Radiological Protection Board 12, 156
nitroglycerine, manufacture of 201–2
Notification of Installations Handling Hazardous Substances (NIHHS) Regulations (1982) 45, 53–4, 55
Nuclear Installations Inspectorate, report of to Sizewell 'B' Inquiry 59–60
nuclear power, energy production by 50, 51
nuclear radiation 36–41
damage from 154–9
major nuclear hazards 157–9
lethal dose of 12
risk analysis in toxicity of 11–12
units of 154
see also Chernobyl disaster, Three Mile Island incident

Pernis explosion 233
Pheasant Wood Inquiry 60–1
phosgene
toxicity of 31, 161
use in methyl isocyanate manufacture 202
pneumoconiosis, deaths due to 20
Potchefstroom disaster 28–9, 122, 159
pressure groups xii, 3
propane, dispersion of 121
propylene
frequency/fatality magnitude data for 93
in Spanish campsite disaster 42–3
Public Inquiries 58–61, 84
Pheasant Wood 60–1
Sizewell 'B' 59–60
public policy and legislation 48–63, 68–9
CIMAH Regulations 54–8
NIHHS Regulations 53–4
policy positions of public authorities 61–2
public inquiries 58–61
UK dual system 51–3

quality control manager, role of 74–5

rail transport incidents 197
Rijnmond disaster 24
Rijnmond Report xi, 5, 6, 30, 41, 87, 122, 174, 185
frequency/fatality magnitude data in 92–5, 186
population composition in 180, 182
risk assessment in 170
risk analyses and perceptions 1–22
acceptable risk 4–6

individual risk 1–4, 16, 157, 158
methodologies for 168–70
population composition 180–2
population density 175–80
presentation of 183–6
public perception of risk 18–21
realism and conservatism 13–15
societal risk 2, 3–4, 15–16, 157, 159
structure for risk assessment 16–18, 56–8
thresholds and uncertainty 6–13
risk contours 2–3, 5, 57, 136, 137, 183–5
individual risk 3, 16
societal risk 3, 15
risk prevention, costs and benefits of 224–36
benefits to employees 229–32
delayed deaths 230–2
fatal accident frequency rate comparisons 230, 231
benefits to society 234–5
capital cost structure 226–9
cost of saving life 232–4
standard costings 224–6
Road Haulage Association, training scheme run by 196
road transport incidents 189, 190, 191–2, 197
causes of 197–8
see also Spanish campsite disaster, Checotah highway explosion
Road Transport Industry Training Board, training scheme run by 196

safety management 64–78
ASME code 70–2
implementation of 72–7
company organization 64–6, 67
compliance with codes, procedures and regulations 68–70
participative control 66–8
process management checklist 77–8
Salford explosion 45, 53
San Juan Ixhuatepec disaster xii, 25, 26–7, 41, 149, 153
Seveso directive 220, 221
Seveso disaster xii, 34–6, 54, 61, 159
Silvertown explosion 24, 229
Sizewell 'B' Inquiry xii, 59–60, 241
failure rate estimates in 97
review of management functions in 76
risk assessment in 3, 14, 56, 157, 158, 159
sodium chlorate
in Salford explosion 45
maximum handling quantity of 53
Spanish campsite disaster xii, 42–3, 188, 189
Spanish edible oil contamination disaster xii
sulphur dioxide, toxicity of 31

2,3,7,8,tetrachlorodibenzoparadioxin (TCDD), in Seveso disaster 34–6
Three Mile Island incident 36
titanium oxide, in Salford explosion 45
TNT equivalent for vapour cloud explosions 104–5, 140, 150–1, 204
toxic gas hazards 27–36, 159–65
effect of shelter 165
inhalation of 6–11, 164
toxic mechanisms 160–1
uncertainty of experimental data 161–3
vulnerable people 164
see also Bhopal disaster, Potchefstroom disaster
transportation of chemicals, risks involved in 41–4, 188–99
comparative transport accident rates 189–91
precautionary measures 193–7
prescription of routes 196
rail transport incidents 197
risks from hazardous road transport 191–2
statistics 190–1, 197–8
transport emergency card (tremcard) 196–7

Warren Springs wind tunnel report 112–18, 122